SPATIAL REASONING
FOR EFFECTIVE GIS

Joseph K. Berry

with a collection of mathematical formulae
by Nigel Waters

Fort Collins, Colorado • 1995

ISBN 978-0-470-23633-8

GIS World Books
155 E. Boardwalk Drive, Suite 250
Fort Collins CO 80525, USA
Phone: 970-223-4848 • Fax: 970-223-5700
E-mail: books@gisworld.com

Edited by John Hughes
Designed and composed by Wade Smith
Project managed by Bea Ferrigno

Second Printing, 1997.

SPATIAL REASONING

Also by Joseph K. Berry:

BEYOND MAPPING:
CONCEPTS, ALGORITHMS AND ISSUES IN GIS

and companion software,

TUTORIAL MAP ANALYSIS PACKAGE (TMAP™)
GIS CONCEPTS DIGITAL SLIDE SHOWS (GCON™)

Contents

Foreword

As a friend of Joe Berry for many years, it is difficult for me to separate my personal respect from professional respect as I write the foreword for his latest book. Having admitted that bias, I congratulate Joe for this excellent work and commend the book to anyone, especially any student, who is as fascinated by the utter primacy of spatial information as I am.

Joe has a unique ability to convey the oftimes complex concepts of spatial data and GIS functions in words that mean something to others less technically versed, and that is important. It is important because an understanding of the analytical capabilities of GIS is crucial to putting the technology to work effectively. There are many benefits to electronic map display but many more to electronic map analysis.

A title of one of Joe's early *GIS WORLD* columns was "As the Crow Walks," which really says it all. This book is replete with Berryisms that amuse as well as teach—Joe's ability to do both is the essence of his charisma. Mark Twain said, "I have never let my schooling interfere with my education." Joe Berry is the Mark Twain of GIS.

H. Dennison Parker
Fort Collins, Colorado
April 1995

Preface

Geographic information systems (GIS) technology has its roots in the traditional map and cartographic process. For thousands of years maps have been used for navigating through unfamiliar terrain and seas. They precisely locate physical features and describe their characteristics and condition. Recently, the analysis of mapped data has become an integral part of the decision-making process in fields as diverse as resource planning, facilities management, environmental modeling, market forecasting and precision farming.

There are numerous similarities and complements between GIS and its traditional mapping heritage. Many GIS capabilities are translations or extensions of time-honored geographic principles. A digital map, however, differs radically from its analog counterpart in that it consists of an organized set of numbers. In some applications the numbers merely serve as surrogates for the inked lines, symbols and colors used to construct a map image. In other applications the numbers are the grist for new procedures involving spatial statistics, map algebra, spatial analysis, and GIS modeling.

This book is about the elements of GIS that make it different from traditional map structure, content, processing, and use. The notion of "spatial reasoning" captures the book's focus. We are at the threshold of a new era—one that directly incorporates the complexity of geographic space in decision making instead of simply applying a single solution throughout an entire area. This step isn't as much rocket science as it is a new approach to problem solving. Sure, there are new and initially confusing tools, but the real challenge is in "thinking spatially."

This book encourages you to extend the traditional role of maps from telling us "Where is what?" to "So what?" It is an invitation to consider the expanded capabilities of GIS and relate them to your current operations. It prompts you to rethink the characteristics of mapped data, your data analysis procedures, and, ultimately, how you use maps. Rarely is the direct automation of existing manual techniques sufficient for today's applications. GIS is a new technology, and as such it presents new opportunities as well as new pitfalls.

This book is an extension of my earlier work, *Beyond Mapping: Concepts, Algorithms, and Issues in GIS* (GIS World Books, 1993) and is similar in style and format. Both are based on a series of articles in *GIS WORLD* magazine's "Beyond Mapping" column, using a lighthearted style and practical examples to convey underlying GIS theory. This book extends, and in some cases elaborates on, the discussion of the map analysis "toolbox" used in GIS modeling and its creative application. Because GIS is a rapidly developing technology, prospective users need to keep up with new developments and approaches. As Will Rogers noted, "Even if you are on the right track, you will get run over if you just sit there." This book is intended to keep you moving beyond basic mapping.

The 10 topics included here provide a broad understanding of the potential and pitfalls of GIS as well as the concepts and procedures of specific map analysis operations. For the most part, these two thrusts are interlaced: Topics 1, 3, 4, 8 and 10 present the broader issues while topics 2, 5, 6, 7 and 9 focus mainly on technical issues. A list of recommended readings is included at the end of each topic and resources for further information appear on pages 195-200. An expansion of Nigel Waters' intriguing discourse on the 10 most beautiful formulae in GIS (*GIS WORLD*, November 1994) begins on page 175. It is a delightful, concise treatise on classical spatial and nonspatial procedures used in GIS. It stresses the logical and conceptual foundations that support mathematical tools and develops an appreciation of their derivation, integration, and continuing evolution. The reader is encouraged to study and enjoy this provocative collection of analytical building blocks.

Two PC-based software packages support the material presented in this book. The Tutorial Map Analysis Package (tMAP™) is a series of hands-on exercises that provide practical experience to reinforce the map analysis concepts presented in each topic. The digital slide shows on basic GIS concepts (gCON™)portray theory and applications that extend many of the figures in this book and more thoroughly develop the graphical explanations of underlying principles. Descriptions of both software packages appear on pages 199-203.

In writing this book, I call on nearly 30 years of GIS experience—some good, some bad. Although the acronym wasn't coined until the mid-1970s, my encounters with maps and computers began in the

1960s at the University of California, Berkeley. Free speech, filthy speech and an occasional course dominated my conscious being, but GIS lurked in my subconscious as an undergraduate research assistant in the Forestry Remote Sensing Lab. After a stint with the Army and a master's degree in business administration, my latent interest in GIS surfaced at Colorado State University's Remote Sensing Program. My doctoral studies, however, were a bit unusual, as my major advisor was a quantum physicist modeling light's interaction in plant canopies.

I mention my patchwork past to establish a couple of simple points. First, all fields have their beginnings and their pioneers. It was my luck to be present at the birth of GIS under the tutelage of great minds and even greater human beings. This book is dedicated to two such visionaries—Drs. Robert N. Colwell and James A. Smith. Bob retired from U.C. Berkeley, but is still the father of remote sensing in the hearts and minds of all who followed him. Jim is a senior scientist with NASA Goddard Space Flight Center and is still at the cutting edge of his field. To his devoted students, however, his support and friendship always will be his greatest contribution.

Second, GIS is an eclectic field, not an isolated discipline. It thrives in a diverse environment that casts forester, geographer, mathematician, marketer, scientist, manager, and a multitude of others into an unlikely mix like stone soup: everyone throws something into the pot —as it simmers, the flavors merge and fortify the broth. Such has been the case with GIS, which is enriched by the varied personalities and perspectives of all who pursue it.

Joseph K. Berry
Fort Collins, Colorado

Introduction

Where is GIS?

Driving Forces, Trends, and Forecasts

Is the GIS Cart in Front of the Horse?

What began in the 1960s as a cartographer's tool has evolved into a revolution in many disciplines. General users have become more directly engaged with GIS technology, radically changing the nature of GIS applications. Early uses emphasized mapping and spatial database management. Increasingly, applications have moved to modeling the interrelationships among mapped variables. Most of these applications have involved *cartographic modeling*, which employs GIS operations to mimic manual map processing techniques, such as map reclassification, overlay, and simple buffering around features. The new wave of applications concentrates on *spatial modeling*, employing spatial statistics and advanced analytical operations. The spatial modeling approaches can be grouped into three broad categories: data mining, predictive modeling, and dynamic simulation.

Data mining uses the GIS to discover relationships among mapped variables. For example, a map of dead and dying spruce/fir timber stands can be statistically compared to maps of driving variables, such as elevation, slope, aspect, soil, and depth to bedrock. If a strong spatial coincidence (correlation) is identified for a certain combination of driving variables, that information can be used to direct management action. In the sickly tree example, if the dead trees tend to be on high, steep, northern slopes with thin, acid soils, then forest managers can ask the GIS to identify areas of trees living in these conditions and take appropriate preemptive action. It's like the remark made by famous robber Willy Sutton when asked why he robbed banks: "That's where the money is." Often the simple relationships hidden in complex data are revealed by a slightly different perspective.

Another form of data mining is the derivation of empirical models. For example, a geographic distribution (3-D surface) of PCB concentrations in an aquifer can be interpolated from water samples taken at several wells. Areas of unusually high concentrations (more than one standard deviation above the average) are isolated. When a time series of the high-concentration maps is animated, the contamination appears to move through the aquifer—hence, an empirical ground water model. A "blob" moving across the map indicates an event, whereas a steady "stream" snaking its way along indicates a continuous discharge of a pollutant.

Most *predictive modeling* is nonspatial. Data are collected by sampling large areas, then reducing the set of measurements to a single typical value (arithmetic average). The average values for several variables are used to solve a mathematical model, such as a regression equation. For example, a prediction equation for the amount of breakage during timber harvesting is defined in terms of percent slope, tree diameter, tree height, tree volume, and percent defect, with big, old, rotten trees on steep slopes having the most breakage. The nonspatial approach ignores the inherent spatial information collected and substitutes the average of each variable into the equation to solve for a single estimate of breakage for an entire area. A GIS solution, however, spatially interpolates the field data into mapped variables, then solves the equation for all locations in space. That approach generates a map of predicted breakage with "pockets" of unusual breakage levels clearly identified. Analogous procedures can detect pockets of unusually high sales of a product, levels of crop productivity, or incidence of disease.

Dynamic simulation allows the user to interact with a spatial model. Model behavior can be investigated by systematically changing the model's parameters and tracking the results. This "sensitivity analysis" identifies the relative importance of each mapped variable within the context of its unique geographic setting. In the timber breakage example, the equation itself may be extremely sensitive to steep slopes. In a project area with only gentle slopes of less than 10 percent, however, tree height might be identified as the most important factor.

A less disciplined use of dynamic simulation enables a GIS to act like a spatial spreadsheet and address "what if" questions. For example, the avoidance of steep slopes and visual connectivity to houses might be considered in a highway siting model. What if steep slopes are considered more important? Where does the proposed route change, and where does it not change? What if visual connectivity is considered more important? This informal use of dynamic simulation actively involves decision makers and interested parties in the map analysis process. The induced dialogue develops a common understanding that greatly exceeds the information packed in a static data sandwich of maps.

What is the reality of these futuristic tools? In many respects, the new applications might have "the cart in front of the horse." GIS can store tremendous volumes of descriptive data and overlay myriad maps for their coincidence. It has powerful tools for expressing the spatial interactions among mapped variables. There is, however, a chasm between GIS and applied science. The reality is that the bulk of our scientific knowledge lacks the spatial specificity in the relationships among variables demanded by these advanced applications. We have a tool that characterizes spatial relationships (*cart*); we lack the research and understanding of its expression in complex systems (*horse*).

For example, a GIS can characterize the change in a relative amount of edge in a landscape by computing a set of fractal dimension maps as a forest is modified. That and more than 20 other landscape analysis indices allow us to track landscape dynamics, but what these changes mean for nesting birds is beyond our current scientific knowledge. Similarly, a GIS can characterize the effective sediment-loading distance from streams as a function of slope, vegetative cover, and soil type. It's common sense that areas with stable soil on gentle, densely vegetated intervening slopes are farther away from a stream (in terms of sediment-loading potential) than areas with unstable soils and steep, sparsely vegetated intervening slopes.

But how are effective sediment-loading distances translated into fish survival? Exactly where can a developer dig up the dirt and not have the dirt balls rain down on the fish? Similarly, neighborhood variability statistics allow us to track the diversity, interspersion and juxtaposition of vegetative cover types. How then are these statistics translated into management decisions about wildlife populations? Exactly where can a logger cut trees without destroying the last spotted owl? These (and many others) are serious questions that can't be solved by technology or science alone.

The ability of GIS to integrate multiple phenomena is well-established. The functionality needed to relate the spatial relationships among mapped variables is in place. What's lacking is the scientific knowledge to exploit these capabilities. Until recently, GIS was thought of as a manager's technology focused on inventory and record-keeping. Even the early scientific applications merely used it as an electronic planimeter to aggregate data over large areas for input into traditional, nonspatial models. I hope we are embarking on an era of scientific research in which spatial analysis plays an integral part and expresses its results in GIS modeling terms. The opportunity to have the scientific and managerial communities use the same technology is unprecedented. Until then, however, foresighted, yet frustrated, managers will be forced to use the analytical power of GIS to construct their own models—based on their common (and occasionally uncommon) sense.

A New Spatial Paradigm

Wha'cha mean a pair-o-dimes; heck, I don't
even have two nickels to rub together.

The mechanics of GIS were made a lot easier in the last decade, but the relationships and assumptions built into GIS models remain mere sketches of uncharted intellectual terrain. Such is the challenge GIS presents to basic and applied science.

But what about the rest of us? Aren't the scientists going to do it all, leaving us to merely click on the right icon? What does the evolving technology have in store for the general user? In short, it offers a mind-expanding (or quite possibly, mind-exploding) paradigm shift in how we perceive, handle and employ maps. We are shifting from a product-focus to a utility-focus in our map dealings. No longer is it what a map contains, but how "that map combined with this map and eye of newt can produce what we really need." That takes us beyond mapping to *spatial reasoning*, meaning that the process and procedures of manipulating maps transcend the mechanics of GIS interaction. The ability to think spatially becomes as important as "How do I do that?"

Most GIS users have cognitive skills reflecting their experiences (both good and bad) with manual map processing and procedures. Their data analysis experience has been with nonspatial data, or measurements in which the spatial component was removed surgically, leaving only an average value. But GIS offers a host of new tools for analyzing mapped data. It follows that these new tools will spawn a new way of doing business with maps, beyond the vocational mastery of a system's user interface.

Cornerstone to this new perspective is an appreciation that maps are data—numbers first, pictures later. That is a radical departure from our 8,000-year history of mapping. In the past, maps primarily were *descriptive*. They showed the precise placement of physical features, usually for navigation purposes. Increasingly, maps have become *prescriptive*, serving as data in determining appropriate management actions. They tell us where it is (inventory), and they provide insight into how it could be (analysis).

Map analysis is an emerging discipline, recognizing fundamental map analysis operations independent of specific applications. These analytical tools extend mapping and management of spatial data to GIS modeling, expressing relationships within and among mapped data. Familiarity with this map analysis toolbox is the initial step toward spatial reasoning. Bizarre concepts, such as a Standard Normal Variable surface or Coefficient of Variation map, must become second nature before you can maneuver your souped-up GIS like a race car driver.

Spatial reasoning is the effective application of these tools to solve problems. That involves developing an understanding of their appropriate use for particular applications. Conceptualization of an elk habitat model, for example, requires an understanding of the important factors (mapped variables) and how to express their interaction (tools) to identify areas of suitable habitat (GIS solution). The GIS specialist, wildlife biologist and natural resource manager must work closely at the onset of model development. The problem can't be dissected into separate pieces and solved independently. Their collective strength lies in the ability to communicate various perspectives of the problem and its comprehensive solution. Because elk habitat is inherently a spatial problem, spatial reasoning becomes the medium of intellectual exchange.

As the civil rights adage says, "The people with the problems are the people with the solutions." Thus, the GIS user must be involved in developing GIS solutions. You can't abdicate the responsibility of GIS model development to someone who just happens to know how to boot-up the GIS black box. Nor can you abandon the years of indigenous knowledge about the unique character of your area to a scientist in another region. Your obligations extend even beyond spatial reasoning skills to *spatial dialog* deftness.

During that final step, GIS is used as a decision support system. The focus is on consensus building and conflict resolution among interested parties. The GIS is used as a means to respond to a series of "what if" scenarios in which any single map solution isn't important. It's how maps change as different perspectives are tried that develops enough information to support a decision. That process includes an understanding of the sensitivities involved in the decision. It also involves decision makers in the analysis process instead of just choosing among a set of tacit decisions (static alternatives) produced by detached analysis. Using GIS in this manner is a radical departure from current spatial reasoning and dialog methodologies. When you reach this plateau, you and your high-powered GIS are ready for the races.

One more point needs to be made. GIS is moving rapidly from the domain of the GIS specialist to the general user, and is about to face the utilitarian user who lacks the sentimental attachment of the earlier GIS zealots. In the past, end users were content with automating existing manual processing and data retrieval systems. As they become more knowledgeable and adept in map analysis and modeling, we can expect increasing demands on GIS that will push at the envelope of traditional concepts of map content, structure and use. Maps will be viewed less as descriptive images and more as mapped data expressing user understanding of spatial interactions. Effective communication of the spatial reasoning that supports modeled maps will become as important as statements of map scale and projection.

What a change of events. The technologists have been pushing GIS for years—both its virtues and products. We are at a point in time when things are about to reverse, and that may be the technologist's worst nightmare. Instead of a small, friendly user community gratefully accepting the bug fixes and new features in each software update, there is a growing community tugging at the existing set of GIS capabilities and applications. The growing cadre of enlightened users are pushing GIS into new areas the technologists never dreamed of. But it's the only way to get your nickel's worth out of GIS...maybe even a new pair of dimes.

Recommended Reading

Books

Berry, J.K. "Epilog" in *Beyond Mapping: Concepts, Algorithms, and Issues in GIS*, Fort Collins, CO: GIS World Books, 1993.

GIS World. "GIS Technology Trends." Chapt. 6 in *1994 International GIS Sourcebook*, Fort Collins, CO: GIS World Books, 1993.

GIS World. "GIS Technology Trends." Chapt 4 in *GIS World Sourcebook 1995*, Fort Collins, CO: GIS World Books, 1994.

Maguire, D.J., and J. Dangermond. "Future GIS Technology." In *The GGI Source Book for Geographic Information Systems*, 113-20. London, UK: The Association for Geographic Information, 1994.

Newell, D. "Where is GIS Technology Going?" In *The GGI Source Book for Geographic Information Systems*, 19-24. London, UK: The Association for Geographic Information, 1994.

Rix, D. "Recent Trends in GIS Technology." In *The GGI Source Book for Geographic Information Systems*, 25-29. London, UK: The Association for Geographic Information, 1994.

Journal Articles

Berry, J.K. "Learning Computer-Assisted Map Analysis." *Journal of Forestry* 39-43 (October 1986).

Berry, J.K. "A Brief History and Probable Future of GIS in Natural Resources." *The Compiler* 12(1): 8-10 (1994).

Boyle, A.R. "Concerns About the Present Applications of Computer-Assisted Cartography." *Cartographica* 18: 31-33 (1981).

Ottens, H. "Relevant Trends for Geographic Information Handling." *Geo-Info Systems* 4(8): 23 (1994).

UNDERSTANDING GIS

HIGH TECHNOLOGY FOR
MID-LEVEL MANAGEMENT

What we are lacking is not facts, but meaning.
—*Ellen Goodman*

As GIS moves from graphical inventories to spatial reasoning, new procedures need to be developed for communicating the logic that supports GIS models. An
end-user needs to interact with a model, a spatial
spreadsheet, which can present alternative perspectives. This section describes the interactive use of a
map pedigree which links GIS commands to a flowchart
of model logic.

Creative Computation

Distinguishing Data from Information and Understanding

1

The digital map is at the core of GIS. From that perspective, a map isn't simply a comfortable, colorful image, but an organized set of numbers. Analyzing these data involves processing thousands upon thousands of numbers to produce a new set of numbers (i.e., a new map). That is a tedious task for humans, so we rely on our sickly gray, silicon friends. It isn't that we couldn't do what computers do; it's just that we have more creative things to do. We are creative (fire), while our computers are computational (ice).

As with any new technology, GIS's sharp contrast acts like an ink-blot test—in one instance it appears to be one thing (image); in the next it's another (numbers). What's needed is a blending of the two perspectives into a middle ground of creative computation. So, what's holding us back? Two things come to mind: (1) the complex nature of spatial problems and (2) the inhumane nature of GIS.

Let's consider the complex nature of spatial problems. Historically, mapping was simply a matter of not getting lost, and maps were used to identify the placement of features on pocket-sized abstractions of our landscapes and seascapes. Then maps evolved into graphical inventories linking "features to attributes" and describing the character, content, and condition of mapped entities. Now we are enamored with the potential of a GIS to address complex spatial problems using "map-emati-cal modeling." From determining the optimal route for a highway to identifying the ideal habitat for spotted owls, GIS is viewed as the decision maker's salvation. But is it really?

It's generally accepted that good data are the prerequisite of good decisions. With the advent of the computer, good data often are equated to voluminous data—the more the better. In reality, mounds of data must be sieved for a subset that's significant and relevant to the decision at hand. The true effectiveness of any information system lies in its ability to distill *data* (all facts) into *information* (useful facts). GIS is adept at swallowing tremendous amounts of spatial data, then repackaging and presenting germane information to the user.

Yet descriptive information isn't enough in many decisions. Consider figure 1.1, which is based on sociology's enduring quandary of fact/value conflicts. The lower right panel of the matrix identifies the ideal condition (COMPUTATIONAL) in which there is social agreement on the facts and values surrounding a decision. That is computer heaven in

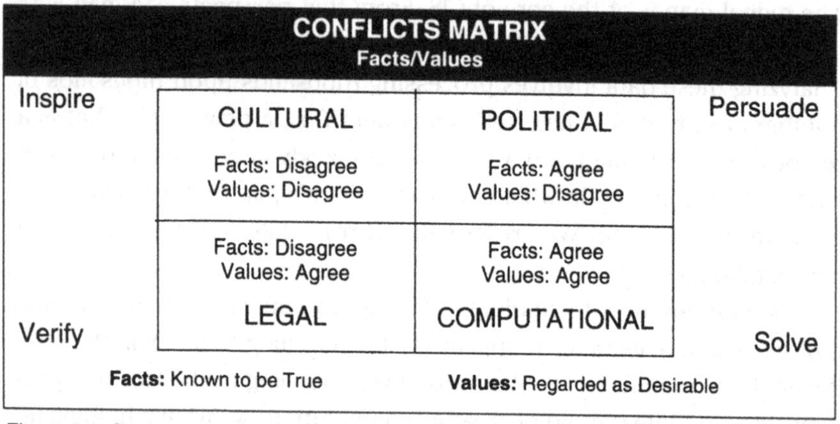

CONFLICTS MATRIX Facts/Values			
Inspire	CULTURAL Facts: Disagree Values: Disagree	POLITICAL Facts: Agree Values: Disagree	Persuade
Verify	Facts: Disagree Values: Agree LEGAL	Facts: Agree Values: Agree COMPUTATIONAL	Solve
	Facts: Known to be True	**Values:** Regarded as Desirable	

Figure 1.1. Only a subset of all land-use issues involves computational solutions. Most use the computer as a vehicle for communicating various perspectives of an issue through spatial reasoning and dialog.

which efforts are focused on perfecting the computational solution. In that role, GIS is viewed as a Decision Making System (DMS), which is coupled tightly to mathematical models. Under these conditions, science and technology are indisputably paramount, with their solutions directly translating into decisions.

However, many of the decisions facing GIS fall outside of the COMPUTATIONAL panel. Consider the LEGAL panel in the lower left in which there is agreement on values, but disagreement on facts. It's like the accused claiming he was at home in bed at the time of the murder, while the prosecutor claims he was at the scene of the crime. In that role, technology is called on to "verify" the facts by viewing and comparing alternative definitions of fact. In techy-speak, that means "establishing the sidebars of the system's response."

A POLITICAL conflict (upper right panel) is the opposite. In that decision environment, there is agreement on facts, but disagreement on values. We might agree on the fact that a species is endangered, but disagree on the relative value we place on environmental and economic considerations. In that role, technology is used to persuade society (or at least an effective majority) of the logical reasoning supporting a position. The CULTURAL panel (upper left) is the murkiest yet—disagreement on facts and values (e.g., the abortion issue). Under these conditions, science and technology are ineffective, with their solutions having little relevance to either the discussion or a decision.

But what does all that esoteric stuff have to do with a GIS? It's just a data sponge that draws awesome graphics, right? Actually it is a Decision Support System (DSS) that transforms data into information. And if used creatively, it can transform information into an *understanding* of the complex nature of spatial problems, which, in turn, can lead to viable decisions. It's preposterous to assume that one more decimal place of accuracy in a spatial model could solve a complex problem in which the facts and/or values are in dispute. The computational approach only works on a limited set of spatial problems.

A view of GIS beyond data and information is in order. A decision maker's understanding is as important to a good decision as good data. To paraphrase Professor Robert Woolsey of the Colorado School of Mines, "Managers would rather live with a problem they can't solve than apply a solution they don't understand." As problems become increasingly gray, the black-and-white approach of computational solutions becomes increasingly limited. The next chapter investigates how GIS and creative computing can be used to extend data to information and, ultimately, to understanding complex spatial problems. What do you say— pipe dream or reality?

Toward a Humane GIS

An Interactive Link Between
Model Logic and GIS Code

2

Chapter 1 noted two things holding back GIS: (1) the complex nature of spatial problems and (2) the inhumane nature of GIS. Many applications go beyond repackaging mapped data into spatial information that is presented to decision makers. Mapematical models relate spatial variables and, in some instances, can be used to "solve" land use issues. However, more often than not, a computational solution isn't possible because complex issues often are driven by conflicts in facts and values among individuals. In these cases words like "verify," "persuade," and "inspire" replace "solve." What is needed in these situations is a decision-making environment that promotes enlightened communication of the impacts of varying fact/value perceptions.

Or, what is needed is a kinder, gentler GIS—one that fully engages decision makers in the spatial analysis process, encourages them to try different interpretations of a spatial model and compare the outcomes, and sanctions spatial reasoning and dialogue. We have the computer, the database, the analytical operations, and even a colorful set of point-and-click icons. But we're missing a succinct expression of a model's logic and an interactive mechanism to execute the model under various interpretations.

Yet don't despair. Your vendor's GIS (Guaranteed Income Stream) is addressing the situation with a humane interface. Consider figure 2.1, a simple model to determine areas suitable for development as being gently sloped and near roads. The upper portion of the figure is a flowchart that graphically summarizes the model's logic as a series of boxes (maps) and lines (operations). Derived maps of slope and proximity to roads are created from the base maps of Elevation and Roads, respectively. The derived maps are "calibrated" in relative terms of suitability, then combined for an overall suitability map. The lineage, or pedigree, of the final map is succinctly expressed in the graphic.[1]

In a humane GIS interface, the flowchart is linked dynamically to the database and the command macro of the model. If you click on a box, descriptive statistics and/or an image of the map pops up. A click on a line pops up a description of the operation and its "parameterization"

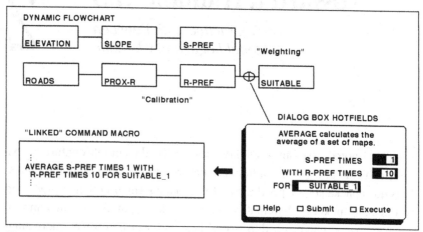

Figure 2.1. Clicking on a line (operation) in the dynamic flowchart pops up a dialog box with hotfields that allow the user to automatically edit and execute the linked command macro.

(lower right portion of the figure). By clicking around the flowchart, you can get a fairly good handle on the logic of the model—sort of a personalized tour of the black box. That exploratory interaction with the model develops an understanding of the spatial reasoning supporting the application. Questions, inconsistencies and gaps in logic are discussed with the model developer. That dialogue can enhance the decision maker's confidence in the model and refine the model as well.

If the user is authorized, another level of interaction can take place. When an operation's dialog box appears by clicking on a line, its specifications are contained in *hotfields*. For example, if the *weighting* step is clicked as shown in the figure, the AVERAGE dialog box pops up and the current weights for the slope (S-PREF) and road proximity (R-PREF) preferences are depicted. Clicking on the Help box provides a more detailed description of the operation and its options. Armed with this knowledge, you can change the weights so the average is more heavily influenced by the preference to be near roads (one to 10 times the influence, as shown). To keep things straight, edit the final map name to SUITABLE_1, as shown.

When you finish editing the hotfields and submit the dialog box, the GIS code associated with that step is edited automatically (lower left portion of the figure). You can pop-up and edit another step in the model if you choose. When you click on the Execute box, your revised command macro is run and the new SUITABLE_1 map is generated. The revised flowchart and macro form the pedigree of the new map, which in turn can be interactively explored and revised. Each time a new scenario is tried, the graphic and coded pedigree serves as the conceptual audit associated with the alternative—an objective record of its fact/value interpretations. Now we're talking.

Comparing the results of your tinkering (SUITABLE_* maps) provides insight into the model's sensitivity and plenty of material for discussion. "Huh, you mean when road proximity is considered 10 times more important, it doesn't change suitability much? And that the change is concentrated in the southwest portion of the project area? Heck, it doesn't affect my property. I thought it would affect every square inch." That direct interaction fully engages stakeholders and decision makers in the analytical process. As they see the effects of their "what if" questions, they develop a better understanding of the issue and its ramifications.

"Naive. Polyannaish. Detached ramblings of a demented soul who never has been in the decision-making trenches. It never will work." These may be a few of the thoughts that cross your mind. Yet consider the alternative: a detached GIS producing an increasing deluge of colorful, yet indecipherable, mapped gibberish. The humane GIS modeling structure provides a user-friendly means for interacting with the GIS that goes beyond map viewing. So what's in it for the GIS specialist? That's for the next chapter.

1. For a more thorough discussion of spatial modeling, see Topic 10 and the Epilog in *Beyond Mapping: Concepts, Algorithms, and Issues in GIS,* GIS World Books 1993, Berry, or "Beyond Mapping," *GIS WORLD*, February-May 1993.

The GIS Modeler's Toolkit

An Object-Oriented Programming System Approach to GIS Model Development

3

So what does the future have in store for the GIS specialist's bag of tricks? Chapter 2 proposed a graphical entry mechanism into GIS application models for users, based on a dynamic flowchart of the final map's pedigree. Such a system allows decision makers to interactively tinker with a model and gain valuable insight into its application—sort of a point-and-click tour of the black box's logical reasoning. If the user is authorized, that tinkering can expand to modifying the weights and calibrations (model parameterization) to generate maps of alternative scenarios. How the final map changes under different interpretations becomes the real spatial information for decision making. The maps themselves are colorful products, but how they change leads to colorful dialogue.

OK, but what does all that have to do with GIS modelers? The techy crowd merely codes, right? Yes, but the link between the command macro and the dynamic flowchart is essential. So why not have the modeler work with the same graphical interface to build the model? Yep, that's it—a graphical, object-oriented, GIS modeler's toolkit. At that level, construction and model structure editing is provided for the modeler, as well as the simple viewing and parameter editing granted to the users.

To get a feel for how it might work, consider figure 3.1. Keep in mind that this approach is somewhere between "scareware" and "vaporware," but it's a probable future for GIS. The top portion contains a set of processing widgets that are tied to data and operations. The lower portion is a graphical workspace. A model developer uses the widgets to construct a flowchart of an application. As the flowchart is completed, the dynamically linked command macro is written automatically.

For example, the modeler might click and drag the box icon, representing a base map, into the work space. Once positioned, the box is defined as the Elevation map. The system searches the database and associates the mapped data with that element of the flowchart (Step 1). Clicking on the box at any time pops up a description and/or display of the map. To continue flowchart construction, the modeler drags the

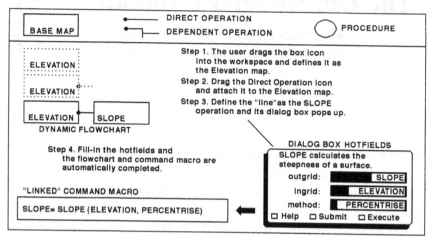

Figure 3.1. The GIS toolkit of the future may allow the modeler to generate fully indexed macros by simply constructing an application's flowchart.

Direct Operation icon and attaches it to the Elevation map (Step 2). At that point, the system knows a new map is desired, but it doesn't know what operation to use.

The modeler defines the line as the SLOPE command, which causes the system to draw an empty output map box and pop up SLOPE's dialog box (Step 3). The example indicates the options for Environmental Systems Research Institute Inc.'s ARC/INFO GIS GRID command. The hot-fields are specified for "out-grid" and "method." (It already knows the "in-grid," unless you want to change it.) The completed dialog box is submitted, which causes the command line to be written to the macro and the out-grid name placed in the flowchart (Step 4). The process is repeated to construct subsequent parts of the model.

Note that not all widgets are the same. Some are related to *data* (maps), some to methods (GIS commands) and their *properties* (command options), while others can identify procedures (frequently used submodels). These terms are in the realm of what computer scientists call *object-oriented programming*.[1] It suffices to say that the basic structure for a humane GIS is in place, but there is a lot of work and discussion before it arrives on your desk.

Continuing with the example model's construction,[2] a modeler might attach and identify the RECLASS operation to the SLOPE map. Complete its hotfields by assigning preferences to various slope classes and designate S-PREF as the outgrid. In GRID syntax, a "remap" table is generated and linked to the RECLASS command line. Other systems will write the reclassification assignments as part of the command line.

That trivial point, however, raises a more important point. The GIS industry has made great strides in establishing standards for data exchange. It wasn't too many years ago that maps in one system couldn't be fed to another. If your temples aren't gray, you probably can't imagine such a silly state of affairs. A database Tower of Babble severely limited GIS's potential.

But what about system-specific application models? It's part of the GIS mystique, isn't it? Part of the insider knowledge that makes your GIS union card so valuable. Part of the product differentiation strategy that ensures system loyalty. It's also part of the evolutionary progression toward open systems.[3] Just as database management systems technology evolved Standard Query Language (SQL) to give a similar look and feel to their users, a GIS Analysis Language (GAL) is the next step in GIS technology.

Because GIS is graphical anyway, why not use a dynamic flowchart of a map's pedigree as the entry point? To users and modelers alike, the flow chart entry would be similar among systems, even if the attached command lines were radically different. That will require an extraordinary effort, similar to the development of geographic standards for the paper map and the exchange standards for the digital map. Yet the payoff is huge. Tackling GIS at the command-line level is a superhuman effort. Keep in mind that the millions of potential GISers are real people who simply need a humane GIS.

1. A complete discussion of object-oriented programming is beyond the scope of this chapter. For more information, see "Object-Oriented Approaches Add Value to GIS," *GIS WORLD*, November 1993.

2. For the complete model, see "Beyond Mapping," *GIS WORLD*, February 1993.

3. For more information, see "Open Systems Mean 'Open Sesame' for the GIS Community," *GIS WORLD*, November 1993.

Recommended Reading

Books

Berry, J.K. "What GIS Is and Is Not: Spatial Data Mapping, Management, Modeling, and More." Topic 4 in *Beyond Mapping: Concepts, Algorithms, and Issues in GIS*, Fort Collins, CO: GIS World Books, 1993.

Berry, J.K. "Implications of a Humane GIS." In *The GGI Source Book for Geographic Information Systems*, 57-62. London, UK: The Association for Geographic Information, 1994.

Medyckyj-Scott, D., and H.M. Hearnshaw (eds.). "Introduction," "The Geographical Information System User," and "Human-Computer Interactions in Geographical Information Systems," in *Human Factors in Geographical Information Systems*, London, UK: Belhaven Press, 1993.

Monmonier, M. "Introduction," "Elements of the Map," and Map Generalization," in *How to Lie with Maps*, Chicago, IL: The University of Chicago Press, 1991.

Journal Articles

Davies, C., and D. Medyckyj-Scott. "GIS Usability: Recommendations Based on the User's View." *International Journal of Geographical Information Systems* 8: 175-89 (1994).

Egenhofer, M.J. "Why not SQL!" *International Journal of Geographical Information Systems* 6(2): 71-85 (1992).

Leung, Y. and K.S. Leung. "An Intelligent Expert System Shell for Knowledge-Based Geographical Information Systems: 1. The Tools." *International Journal of Geographical Information Systems* 7(3): 189-99 (1993).

Yeung, Y. and K.S. Leung. "An Intelligent Expert System Shell for Knowledge-Based Geographical Information Systems: 2. Some Applications." *International Journal of Geographical Information Systems* 7(3): 201-13 (1993).

Wang, F. "Towards a Natural Language User Interface: An Approach of Fuzzy Query." *International Journal of Geographical Information Systems* 8: 143-62 (1994).

FROM FIELD SAMPLES TO MAPPED DATA

ASSESSING GEOGRAPHIC DISTRIBUTIONS

Men have been drawing maps and so studying spatial
patterns for millennia . . . the need to reduce such
information to numbers is rather recent.
— *Brian Ripley*

In the simplest sense, statistics is merely a collection of
numbers. Traditional statistical analysis characterizes
the "typical response" in a data set, whereas spatial statistics seeks to map the data's pattern in geographic
space. This section compares the two approaches and
investigates various techniques of spatial interpolation.

Averages are Mean

Comparing Nonspatial and Spatial
Distributions of Field Data

4

Remember your first brush with statistics? The thought likely conjures up a prune-faced mathematics teacher who determined the average weight of the students in your class. You added the students' individual weights, then divided by the number of students. The average weight, or more formally stated as the *arithmetic mean* (aptly named for the mathematically impaired), was augmented by another measure termed the *standard deviation*. Simply stated, the mean tells you the typical value in a set of data and the standard deviation tells you how typical that typical is.

So what does that have to do with GIS? It's just a bunch of maps accurately locating physical features on handy, fold-up sheets of paper or colorful wall-hangings. Right? Actually, GIS is taking us beyond mapping—from images to mapped data that are ripe for old prune-face's techniques.

Imagine your old classroom with the bulky football team in the back, the diminutive cheerleaders in front and the rest caught in between. Two things (at least, depending on your nostalgic memory) should come to mind. First, all of the students didn't have the same typical weight: some were heavier and some were lighter. Second, the differences from the typical weight might have followed a geographic pattern: heavy in the back, light in the front. The second point is the focus of Topic 2's chapters—describing *spatial statistics*, which map the variation in geographic data sets.

Figure 4.1 illustrates the spatial and nonspatial character of a set of animal activity data. The right side of the figure lists the data collected at 16 locations (Sample #1-16) for two 24-hour periods (P1 in June and P2 in August). Note the varying levels of activity—0 to 42 for Period 1 and 0 to 87 for Period 2. Because humans can't handle more than a couple of numbers at a time, we reduce the long listings to their average value—19 for Period 1 and 23 for Period 2. We quickly assimilate these findings, then determine whether the implied activity change from Period 1 to Period 2 is too little, too much, or just right (like Goldilocks' assessments of Mama, Papa, and Baby Bears' things). Armed with that knowledge, we make a management decision, such as "blow 'em away," or, in politically correct wildlife-speak,

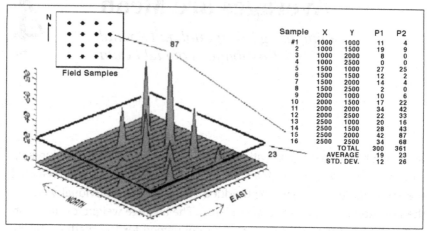

Sample	X	Y	P1	P2
#1	1000	1000	11	4
2	1000	1500	19	9
3	1000	2000	8	0
4	1000	2500	0	0
5	1500	1000	27	25
6	1500	1500	12	2
7	1500	2000	14	4
8	1500	2500	2	0
9	2000	1000	10	6
10	2000	1500	17	22
11	2000	2000	34	42
12	2000	2500	22	33
13	2500	1000	20	16
14	2500	1500	28	43
15	2500	2000	42	87
16	2500	2500	34	68
		TOTAL	300	361
		AVERAGE	19	23
		STD. DEV.	12	26

Figure. 4.1. Spatial comparison of Field Samples and their arithmetic mean.

"hold a special hunt to 'cull' the herd for its own good." It's obvious that animal activity is increasing at an alarming rate throughout the project area.

Whoa! You can't say that, and the nonspatial statistics tell you so. That's the role of the standard deviation. A general rule (termed the *coefficient of variation* for the techy-types) tells us, "If the standard deviation is relatively large compared to the arithmetic average, you can't use the average to make decisions." Heck, it's bigger for Period 2! These numbers are screaming "warning, hazardous to your (professional) health" if you use them in a management decision. There is too much variation in the data; therefore, the computed typical isn't very typical.

Solutions in Geographic Space

So what's a wildlife manager to do? A simple solution is to avoid pressing the calculator's standard deviation button, because all it seems to do is trash your day. And we all know that complicated statistics stuff is just smoke and mirrors, with weasel-words like "likelihood" and "probability." Bah, humbug. Go with your gut feeling (and hope you're not asked to explain it in court).

Another approach is to take things a step further. Maybe some of the variation in animal activity forms a pattern in geographic space. What do you think? That's where the left side of the figure comes in. We use the X,Y coordinates of the Field Samples to locate them in geographic space. The 3-D plot shows the geographic positioning (X=East, Y=North) and the measured activity levels (Z=Activity) for Period 2. I'll bet your eye is "mapping the variation in the data"—high activity in the Northeast, low in the Northwest, and moderate elsewhere. That's real information for an on-the-ground manager.

The thick line in the plot outlines the plane (at 23) that spatially characterizes the "typical" animal activity for Period 2. The techy-types might note that we often split hairs in characterizing this estimate (fitting Gaussian, binomial or nonparametric density functions), but in the end, all nonspatial techniques assume that the typical response is distributed uniformly in geographic space. For nontechies, that means the different techniques might shift the average more or less, but whatever it is, it's assumed to be the same throughout the project area.

But your eye tells you that guessing 23 around Sample #15, where you measured 87 and its neighbors are all above 40, is likely an understatement. Similarly, a guess of 23 around Sample #4, where both it and its neighbors are 0, is likely an overstatement. That's what the relatively large standard deviation was telling you: Guess 23, but expect to be way off (+26) a lot of the time. However, it didn't give you any hints as to where you might be guessing low and where you might be guessing high. It couldn't, because the analysis is performed in numeric space, not the geographic space of a GIS.

That's the main difference between classical statistics and spatial statistics: Classical statistics seeks the central tendency (average) of data in numeric space; spatial statistics seeks to map the variation (standard deviation) of data in geographic space. That is an oversimplification, but it sets the stage for further discussions of spatial interpolation techniques, characterizing uncertainty and mapematics. Yes, averages are mean, and it's time for kinder, gentler statistics for real-world applications.

Surf's Up

*Fitting Continuous Map Surfaces
to Geographic Data Distributions*

5

Chapter 4 introduced the fundamental concepts behind the emerging field of spatial statistics. It discussed how a lot of information about the variability in field-collected data is lost using conventional data analysis procedures. Nonspatial statistics accurately reports the typical measurement (arithmetic mean) in a data set and assesses how typical that typical is (standard deviation). However, it fails to provide guidance as to *where* the typical is likely too low and *where* it's likely too high. That's the jurisdiction of spatial statistics and its surface modeling capabilities.

The discussion compared a map of data collected on animal activity to its arithmetic mean, similar to inset (a) in figure 5.1. The 16 measurements of animal activity are depicted as floating balls, with their relative heights indicating the number of animals encountered in a 24-hour period. Several sample locations in the northwest recorded zero animals, while the highest readings of 87 and 68 are in the northeast (see Chapter 4 for the data set listing).

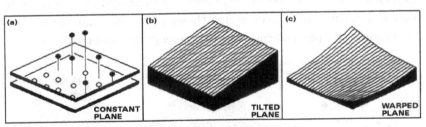

Figure 5.1. Surface-fitted approximations of geographic distribution.

The average of 23 animals is depicted as the floating plane, which balances the balls above it (dark) and the balls below it (light). In techy-speak, and using much poetic license, "It is the best-fitted horizontal surface that minimizes the squared deviations [from the plane to each floating ball]." In conceptual terms, imagine sliding a window pane between the balls so you split the group in half. OK, so much for the spatial characterization of the arithmetic mean (that's the easy stuff).

Now relax the assumption that the plane has to remain horizontal (figure 5.1b). Tilt it every which way until you think it fits the floating balls even better (the squared deviations should be smaller). Being a higher life form, you might be able to conceptually fit a tilted plane to a bunch of floating balls, but how does the computer mathematically fit a plane to the pile of numbers it sees? In techy-speak, "It fits a first degree polynomial to two independent variables." To the rest of us, it solves for an ugly equation that doesn't have any exponents. How it actually performs that mathematical feat is best left as one of life's great mysteries, such as how they get the ships inside those tiny bottles (see Appendix A's reference on *polynomial surface* fitting if you have to know).

Figure 5.1c relaxes yet another assumption: The surface has to be a flat plane. In that instance, in techy-speak, "it fits a second degree polynomial to two independent variables" by solving for another ugly equation with squared terms—definitely not for the mathematically faint-of-heart. That allows our plane to become pliable and pulled down in the center. If we allowed the equation to get really ugly (Nth degree equations), our pliable plane could get as warped as a flag in the wind.

Figure 5.2 shows a different approach to fitting a surface to our data— *iteratively smoothed*. Consider the top series of insets. Imagine replacing the floating balls (figure 5.2a) with columns of modeler's clay rising to the same height as each ball (figure 5.2b). In effect you have made a first order guess of the animal activity throughout the project area by assigning the closest field sample. In techy terms, you generated Thessian polygons, with sharp boundaries locating the perpendicular bisectors among neighboring samples.

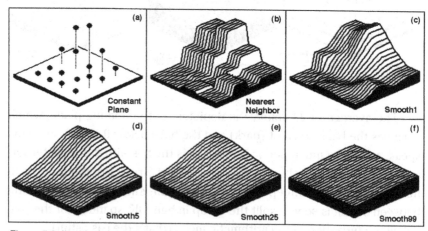

Figure 5.2. Iteratively smoothed approximations of geographic distribution.

Now for the fun stuff. Imagine cutting away some of the clay at the top of the columns and filling in at the bottom. However, your computer can't get in there and whack away like you, so it mimics your fun by moving an averaging window around the matrix of numbers forming the nearest neighbor map. When the window is centered over one of the sharp boundaries, it has a mixture of big and small map values, resulting in an average somewhere in between—a whack off the top and a fill in at the bottom.

Figure 5.2c shows the results of one complete pass of the smoothing window. The lower set of insets (figure 5.2d-f) show repeated passes of the smoothing window. Like erosion, the mountains (high animal activity) are pulled down and the valleys (low animal activity) are pulled up. If everything goes according to theory, eventually the process approximates a horizontal plane floating at the arithmetic mean.

That brings you back to where you started—23 animals assumed everywhere. So what did all that GIS gibberish accomplish? For one thing, you now have a greater appreciation of the potential and pitfalls in applying classical statistics to problems in geographic space. For another, you have a better feel for a couple of techniques used to characterize the geographic distribution of a data set. (Chapter 6 will take these minutiae to even more lofty heights. ... I bet you can't wait.) More importantly, however, now you know surf's up with your data sets.

The Sticky Floor

Inverse Distance, Kriging, and Minimum Curvature Techniques

6

Have you heard of the glass ceiling in organizational structure? In today's workplace there is an even more insidious pitfall holding you back: the sticky floor of technology. You are assaulted by cyber-speak and new ways of doing things. You might be good at what you do, but "they" keep *changing what you do.*

For example, GIS takes the comfortable world of mapping and data management into a surrealistic world of mapematics and spatial statistics. Chapters 4 and 5 described how *discrete field data* can be used to generate a *continuous map surface* of the data. These surfaces extend the familiar (albeit distasteful) concept of central tendency to a map of the geographic distribution of the data. Whereas classical statistics identifies the typical value in a data set, spatial statistics identifies *where* you might expect to find unusual responses.

Chapter 5 described *polynomial fitting* and *iterative smoothing* techniques for generating map surfaces. Now let's tackle a few more. But first, let's look for similarities among the various techniques. They all generate estimates of a mapped variable based on the data values within the vicinity of each map location. In effect, that establishes a roving window that moves about an area summarizing the field data it encounters. The summary estimate is assigned to the center of the window, then the window moves on. The extent of the window (both size and shape) sways the result, regardless of the summary technique. In general, a large window capturing a bunch of values tends to smooth the data. A smaller window tends to result in a rougher surface with abrupt transitions.

Three factors affect the window's extent: its reach, number of samples, and balancing. The *reach*, or search radius, sets a limit on how far the computer will go in collecting data values. The *number of samples* establishes how many data values should be used. If there are more than enough values within a specified reach, the computer uses just the closest ones. If there aren't enough values, it uses all it can find within the reach. *Balancing* attempts to eliminate directional bias by ensuring that the values are selected in all directions around the window's center.

Once a window is established, the summary technique comes into play. *Inverse distance* is easy to conceptualize. It estimates a value for an unsampled location as an average of the data values within its vicinity. The average is weighted, so the influence of the surrounding values decreases with the distance from the location being estimated. Because that is a static averaging method, the estimated values never exceed the range of values in the original field data. Also, it tends to pull down peaks and pull up valleys in the data. Inverse distance is best suited for data sets with random samples that are fairly independent of their surrounding locations (i.e., no regional trend).

The left portion of figure 6.1 contains contour and 3-D plots of the inverse distance (squared) surface generated from the animal activity data described in Chapters 4 and 5 (Period 2 with 16 evenly spaced sampled values from 0 to 87). Note that the inverse distance technique is sensitive to sampled locations and tends to put bumps and pock-marks around these areas.

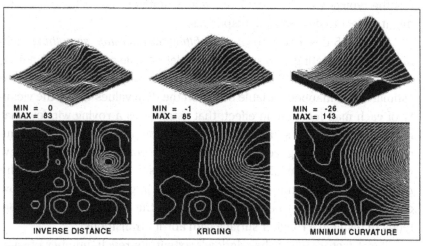

Figure 6.1. Comparison of different interpolation techniques. Interpolation and plots generated by SURFER, Golden Software, Inc.

Opaquely speaking, kriging uses regional variable theory based on an underlying linear variogram. That's techy-speak implying that there is a lot of math behind this one. In effect, the technique develops a custom window based on the trend in the data. Within the window, data values along the trend's direction have more influence than values opposing the trend. The moving average that defines the trend in the data can result in estimated values that exceed the field data's range of values. Also, there can be unexpected results in large areas without data values. The technique is most appropriate for systematically sampled data exhibiting discernible trends.

The center portion of figure 6.1 depicts the *kriging* surface of the animal activity data. In general, it appears somewhat smoother than the inverse method's plot. Note that the high points in the same region of the map tend to be connected as ridges, and the low points are connected as valleys.

Minimum curvature first calculates a set of initial estimates for all map locations based on the sampled data values. Similar to the iterative smoothing technique discussed in Chapter 5, minimum curvature repeatedly applies a smoothing equation to the surface. The smoothing continues until successive changes at each map location are less than a specified maximum absolute deviation, or a maximum number of iterations has been reached. In practice, the process is done on a coarse map grid and repeated for finer and finer grid spacings until the desired grid spacing and smoothness are reached. As with kriging, the estimated values often exceed the range of the original data values and things can go berserk in areas without sample values.

The right portion of figure 6.1 contains the plots for the minimum curvature technique. Note that it's the smoothest of the three plots and displays a strong edge effect along the unsampled border areas.

You likely concede that GIS is sticky, but it also seems a bit fishy. The plots show radically different renderings for the same data set. So, which rendering is best? And how good is it? That's for Chapter 7's discussion of uncertainty mapping.

Recommended Reading

Books

Burrough, P.A. "Methods of Spatial Interpolation," in *Principles of Geographical Information Systems for Land Resources Assessment*, Oxford, UK: Clarendon Press, 1986.

Cressie, Noel. "Spatial Prediction and Kriging" and "Applications of Geostatistics," Chapts. 3 and 4 in *Statistics for Spatial Data*, New York, NY: John Wiley & Sons, 1993.

Fotheringham, S., and P. Rogerson. "Integrating GIS and Spatial Analysis: An Overview of the Issues," Part 1 in *Spatial Analysis and GIS*, London: Taylor & Francis, 1994.

Journal Articles

Okabe, A., B. Boots, and K. Sugihara. "Nearest Neighbourhood Operations with Generalized Voronoi Diagrams: A Review." *International Journal of Geographical Information Systems* 8(1): 43-71 (1994).

Implementing GIS

Considerations, Contingencies, and Confusion

Technology doesn't always give us what we want,
but we're expected to want what it gives.
— *Ellen Goodman*

GIS technology is much more than hardware, software, and data. It begins with a thorough understanding of its intended applications and operating environment. This section presents an applications-driven procedure for assessing GIS information needs within an organization and establishes a hierarchy of questions it can address.

Questioning GIS

Information Needs and
GIS Reality Assessments

7

GIS can answer all of your questions—at least that's what we hear from overzealous marketers. True, GIS has resounding success in many areas, but it also meets with expensive and embarrassing failures in others. Is there any pattern to the technology's successes and failures? What types of applications have high probability of success? Which are doomed to be duds? What conditions affect the likelihood of success? How can you mitigate these conditions?

These are the real questions surrounding GIS; you need to grapple with them before you break the shrink wrap on your new system. The starting point is an Information Needs Assessment (INA), which envisions GIS products, then works backward to derive the intermediate and base maps supporting each product. The process involves four steps:

1. List the application areas to which GIS might contribute.

2. For each application area, describe specific GIS outputs to include a sketch and legend of the final map.

3. For each final map, determine its base maps by successively deriving its supporting maps (with sketch and legend) and the GIS analysis tools needed at each step.

4. Construct two tables summarizing the number of times each base map and each GIS tool are referenced in the various proposed applications.

The INA process is best done with a large group of end users, lightly sprinkled with GIS specialists. The role of the end users is to "blue-sky-it" and envision what they need, not simply to state what they currently produce. The role of the GIS specialist is to stimulate new approaches and assist in deriving the supporting maps and identifying the GIS tools required. In most instances the INA is where the GIS rubber finally meets the application road. It translates GIS rhetoric into the specific context of the organization. Often the INA also can transform into a psychological home run, because it encourages end-user participation and builds a vested interest in GIS at the grassroots level.

As an example of the thought process, consider a map of sensitive areas for resource planning. An end user might envision the map as a set of contiguous polygons that divide a project area into high, medium, and low sensitivity. The user sketches such a map and assigns values 1, 2, and 3, respectively. Next, consider what sort of maps might contribute to the final map. These might include a map of relative visual exposure to roads with areas of high visual exposure identified as high sensitivity.

Now you've identified your first GIS tool (renumber) and supporting map (vexpose). Sketch the map of visual exposure to roads. It is continuous data expressed in raster format, with values increasing from zero (never seen) to large values (seen a lot) assigned to each grid space. So how could you create such a map? You need to RADIATE a map of roads over a map of elevation. This step identifies two input maps (roads and elevation) and another GIS tool (radiate). Sketching these maps focuses attention on the level of detail required. Are different types of roads needed? What about trails? Are there special scenic turnouts? Relate these concerns to the actual numbers that will represent the map features.

Now where are you going to get the maps? Buy them if you can; encode them if you can't. They represent base maps (facts on the landscape), the lowest level of abstraction in a spatial model. You have hit ground zero for this component of mapping sensitivity. What other factors need to be considered—terrain steepness, special habitat areas, proximity to human activity? Repeat the process of distilling the base maps from the conceptual maps for each consideration. Then tackle another GIS product in a similar manner.

When you complete them all (or reach the point of exhaustion), summarize the results. The listing of base maps gives you a handle on database design—which maps are needed, their relative importance, what level of detail, etc. The listing of GIS tools gives you a handle on system design—data loading levels, networking requirements, functionality needed, etc. The "logical fabric" from the distillation of each GIS product gives you a handle on the application modeling effort—type of model, relative degree of difficulty, common intermediate maps, etc. More importantly, the intellectual exercise raises the general awareness of GIS, generates vested interest in its successful implementation, and identifies in-house zealots who will carry the flag.

The results of the INA process directly lead to a second phase of analysis: the development of a GIS Reality Assessment (GRA). That's where the "blue-sky visions" are hammered into a business model commensurate with the organization's resources and culture. The process involves three steps:

1. Develop an implementation scenario for meeting the GIS products identified in the INA process.

2. Determine how and how much it will cost to acquire the data and capabilities implied by the scenario.

3. Repeat steps 1 and 2 until a "realistic" implementation plan emerges.

The GRA process is best done with a working group of managers and a small contingent of GIS specialists. The managers identify the various priorities and tradeoffs among the array of possible GIS products. The GIS specialists tackle the cost of the system, database, and application modeling implied for each scenario. The "realistic" implementation plan should contain a timeline that meets all of the viable GIS products needed—it's just how and how quickly you buy into it. For example, a solution might fully implement one unit (e.g., research), then bring on the other units. Another organization's solution might be to partially implement all units (e.g., inventory capabilities), then expand to more advanced capabilities and applications identified in the GIS products.

An alternative to the INA/GRA process is the "fish-or-cut-bait" scheme of buying a GIS and seeing what happens. Like casting a seed to the wind, if it happens to land in a fertile place a sturdy tree will grow. But keep in mind that nature produces thousands of seeds so just one might flourish.

For related discussions, see "Scoping GIS: What To Consider," in Berry: *Beyond Mapping: Concepts, Algorithms, and Issues in GIS* (GIS World Books, 1993) or "Beyond Mapping," *GIS WORLD*, May-July 1992.

Ask Not What You Can Do for GIS

8

But What GIS Can Do for You!
Seven Basic Questions Addressed by GIS Technology

GIS raises as many questions as it answers—maybe more. As a general rule, the confusion surrounding GIS implementation is inversely proportional to the effort spent in assessing what it can do. Some say it can do everything; some say it can only mess things up. Chapter 7 described a two-part methodology for realistically assessing what GIS can do for you.

An important ingredient of the process is an understanding of the basic questions GIS can answer. Table 8.1 identifies seven questions encompassing most implementations. The questions are ordered progressively from inventory-related (data) to analysis-related (understanding) as identified by their *function* and *approach*.

Table 8.1. There are seven types of questions addressed by GIS. The first three are inventory-related; the latter four are analysis-related and investigate the interrelationships among mapped data beyond simple coincidence.

What GIS Can Do for You		
QUESTIONS FOR GIS	**FUNCTION**	**APPROACH**
1. Can you map that?	Mapping	Inventory
		DATA
2. Where is what?	Management	
3. Where has it changed?	Temporal	
4. What relationships exist?	Spatial	INFORMATION
5. Where is it best?	Suitability	
6. What affects what?	System	
		UNDERSTANDING
7. What if...?	Simulation	Analysis

The most basic question, **Can you map that?**, is where GIS began 30 years ago: automated cartography. A large proportion of GIS applications still involves updating and outputting map products. As an alternative to a room full of draftspersons and rapidograph pens, the digital map is a clear winner. Applications that respond to this question easily are identified in an organization and productivity "payoffs" are apparent.

These mapping applications often are restatements of current inventory-related activities.

Questions involving **Where is what?** exploit the linkage between the digital map and database management technology. These questions are usually restatements of current practices and can get a group to extend their thinking to geographic searches involving coincidence of data they never thought possible. The nature and frequency of such questions provide valuable insight into system design. For example, if most applications require interactive map queries of a corporate database from a dispersed set of offices, you have a major networking headache in store. However, if the queries are localized and turnaround less demanding, a simple sneaker net might suffice.

Where has it changed? questions involve temporal analysis. These questions mark the transition from inventory-related data searches to packaging information for generating plans and policies. Such questions usually come from managers and planners, whereas the questions noted previously support daily operations. A graphic portrayal of changes in geographic space, whether of product sales or parts per million of lead in well water, affords a new perspective on existing data. The concept of "painting" data, which normally are viewed as tables, might initially be a bit uncomfortable. That's where GIS evolves from simply automating current practices to providing new tools.

What relationships exist? questions draw heavily from the GIS toolbox of analytic operations. How far is it from here to there? Can you see the development from over there? How steep is it? Is the cover type more diverse here or there? These are a few examples of that type of question. Whereas the earlier questions involved querying and repackaging base data, spatial relationship questions involve derived information. Uncovering these questions is a bit like the eternal question: Did the chicken or the egg come first? If you don't know what GIS can do differently, chances are you aren't going to ask it to do anything different. How many times have you heard a land use planner say, "I need a weighted visual exposure density surface before we can site the powerline"?

Suitability models spring from questions of **Where is it best?** Often these questions are the end products of planning and are the direct expression of goals and objectives. The problem is that spatial considerations historically are viewed as input to the decision process—not part of the "thruput." Potential GIS users tend to specify the composition (base and derived maps) of "data sandwiches" that adorn the walls during discussion. The idea of using GIS modeling as an active ingredient in the discussion is totally foreign. Suitability questions usually require the gentle coaxing of the INA process described in Chapter 7.

What affects what? questions involve system models—the realm of the scientist and engineer. In a manner of speaking, a system model is like an organic chemist's view of a concoction of interacting substances, whereas a suitability model is analogous to the recipe for a cake. The tracking of "cause and effect" and reliance on empirical relationships are the main ingredients. The same hurdle in identifying these applications exists: the perception that GIS simply provides input. The last 100 years have been spent developing techniques to best aggregate spatial complexity (e.g., stratified random sampling). The idea that GIS modeling retains spatial specificity and responds to spatial autocorrelation of field data is a challenging one.

What if...? questions involve the iterative processing of suitability or system models. For suitability models, they provide an understanding of different perspectives on a project. If visual impact is the most important consideration, where would it be best for development? What if road access is most important? For system models, they provide an understanding of uncertain or special conditions. What would be the surface runoff if there was a 2-inch rainstorm? What if the ground was saturated?

In asking what GIS can do for you, the first impulse is to automate what you do. The stretch to find if any of the other basic questions apply will give you an idea of what GIS can really do for you.

Build It and They Will Come

Technical and Conceptual Considerations in GIS Implementation

9

To many people, GIS is simply a hot technology that should be implemented in every organization. To the more deliberate types, it's a new technology that should be viewed with great suspicion. Regardless of your orientation, it's the implementation phase that makes or breaks GIS in any organization. There are four basic considerations in implementing GIS: hardware/software, database, application models and human impacts (see figure 9.1).

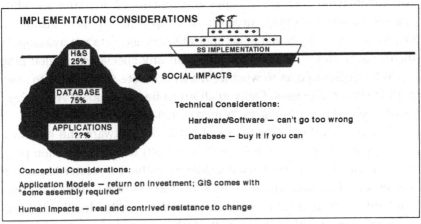

Figure 9.1. There are four basic considerations in implementing GIS: hardware/software, database, application models and human impacts.

Selecting appropriate hardware/software often receives disproportionate attention. In part, the technical aspects provide a comfortable setting for meticulous evaluation—storage capacity, processing speed and real dollars easily are defined. What's often overlooked, however, is the dynamic nature of hardware/software factors. Hardware is in constant flux, and what's considered a technical (or price) barrier today becomes commonplace in tomorrow's boxes. The same holds true for software, as GIS packages continue to leap-frog capabilities with each update.

As a general rule, the larger the organization the more effort is spent on scoping hardware/software. Large government procurements approach "cyber-seizer," because by the time they finally compile a detailed specification, a new generation of technology hits the street. GIS software, however, still commands product loyalty amid a quagmire of different user interfaces. This Tower of Babble has yet to be breached, but the exchange of data is a nonissue. Keep in mind that you can't go too wrong, because when you scrap your computer in a couple of years you can jump to a new GIS package without losing your database; hardware/software is semifluid, but not necessarily quicksand.

Your database, however, is a long-term commitment. Also, it represents the lion's share of the bag of gold necessary to acquire GIS. The best advice is buy it if you can. An increasing amount of mapped data is available in digital form, such as the U.S. Geological Survey's Digital Line Graph (DLG) and Digital Elevation Model (DEM) maps. These data have two advantages: They are cheaper and sanctioned. In-house encoding has such a steep learning curve that it's impractical in most instances. Out-house encoding is viable for special maps and data not available in digital form. However, keep in mind that any specially encoded map could be questioned as to whether it's as accurate as the de-facto standard everyone else uses. Can you afford to be the oddball? A wary eye should be cast upon any *specialty* map nominated for encoding.

The technical considerations of hardware/software and database development usually consume most (if not all) of implementation planning. In reality, the conceptual considerations have a greater impact on successful GIS implementation. As shown in figure 8.1, most GIS costs are hidden and difficult to estimate, with the readily identifiable hardware/software costs just the tip of the iceberg.

Chapters 6 and 7 emphasized that scoping of the GIS products and procedures needed in your organization should drive the implementation process, not system considerations alone. GIS comes with "some assembly required" beyond system setup and database compilation, and model development often sinks the ship.

The development of application models is where GIS's return on investment occurs. The process involves writing *command macros* in the selected GIS language. For the uninitiated, that step seems nearly impossible, with a stack of reference manuals sustaining the steep learning curve. For the cyber-phobiac, the step is impossible and met with fervent resistance.

As with database encoding, you can choose to develop your application models either in-house or out-house. If you choose in-house development, you need to allow for new hires or considerable time for retreading existing personnel. If you choose out-house development, you need to factor in the difficulties in communication, lack of self-determination and a continuous cost stream. Most organizations straddle the issue and hire a consultant to develop a basic set of application models with the active participation of their own GIS specialists in waiting. These on-the-job-training expenses (both in dollars and time) take many GIS planners by surprise. In addition, application models are software specific and increasingly lock you into your GIS package. You can easily flush your platform and transfer your database, but reworking your models into another GIS package is a major undertaking.

Keep in mind that without useful models, the GIS platform and database is like an expensive boat gassed up with high octane fuel, but missing a driver and place to go. Application models provide the utility to a GIS. But even with the best platform, database, and models, you still aren't assured success. The human factor is like a floating mine waiting to sink the ship. If end users see GIS as unfamiliar, overbearing, obtrusive, and threatening, you're doomed from the start. The problem is that's an accurate description of GIS for someone outside the technology. As much attention and concerted effort is needed for developing user acceptance as is paid to the hardware/software and database issues. The social sciences have been wrestling with the human impacts of technology for years. However, most GIS plans pay little more than lip service to these concerns. Traditionally, the technical considerations receive the most attention in GIS implementation planning. But in reality the conceptual considerations are the real determinants of success. Therein lies the weak link in GIS implementation. It's like the field of dreams prophecy: Build it (a GIS) and they (uses and users) will come.

Recommended Reading

Books

Aronoff, S. "Implementing a GIS," Chapt. 8 in *Geographic Information Systems: A Management Perspective*, Ottawa, Canada: WDL Publications, 1989.

Berry, J.K. "Scoping GIS: What To Consider." Topic 8 in *Beyond Mapping: Concepts, Algorithms, and Issues in GIS*, Fort Collins, CO: GIS World Books, 1993.

Burrough, P.A. "Choosing a Geographical Information System." Chapt. 9 in *Principles of Geographical Information Systems for Land Resources Assessment*, Oxford, UK: Clarendon Press, 1986.

Ives, M.J., and K.J. Crawley. "GIS Implementation Issues." In *The AGI Source Book for Geographic Information Systems*, 39-43. London, UK: The Association for Geographic Information, 1994.

Korte, G.B. "Selecting and Implementing a GIS." Part 2 of *The GIS Book* (Third Edition), Santa Fe, NM: OnWord Press, 1993.

Journal Articles

Campbell, H.,and I. Masser. "Implementing GIS: The Organizational Dimension." *Mapping Awareness* 8(2): 20-21 (1994).

Goodchild, M.F., R. Haining, and S. Wise. "Integrating GIS and Spatial Data Analysis: Problems and Possibilities." *International Journal of Geographical Information Systems* 6(5): 407-23 (1992).

Langran, G. "Issues of Implementing a Spatiotemporal System." *International Journal of Geographical Information Systems* 7(4): 305-14 (1993).

Walker, D.R.F., et. al. "A System for Identifying Datasets for GIS Users." *International Journal of Geographical Information Systems* 6(6): 511-27 (1992).

TOWARD AN HONEST GIS

PRACTICAL APPROACHES TO
MAPPING UNCERTAINTY

True genius resides in the capacity for evaluation
of uncertain, hazardous and conflicting information.
— *Sir Winston Churchill*

By their very nature, maps are abstractions of real conditions. They approximate the positioning of tangible or
conceptual features on our landscape with varying
degrees of certainty. This section introduces the concept of a shadow map of certainty and its use in tracking error propagation in GIS models.

The This, That, There Rule

Creating a Shadow Map of Certainty

10

You've heard it before: "This map says we are on *that* mountain over *there*." Yep, maps aren't always perfect regarding precise placement of discernible features. Mark Monmonier, in his insightful book *How to Lie with Maps*, notes that it's "not only easy to lie with maps, it's essential...to present a useful and truthful picture, an accurate map must tell white lies."[1] So how can we sort the little white lies from the more serious ones? Where is a map most accurate; where is it least accurate? If it's not correct, what is the next most likely condition? In short, what does it take to get an honest map?

First, we need to recognize that these white lies are necessary, because maps (1) distort the 3-D world into a 2-D abstraction, (2) selectively characterize just a few elements from the actual complexity of spatial reality, and (3) attempt to portray environmental gradients and conceptual abstractions as distinct spatial objects. The first two concerns have challenged geographers and cartographers since the inception of map-making, and have resulted in at least de-facto standards for most of these issues. The third concern, however, is more recent and involves fuzzy logic and spatial statistics in its expression.

One approach to the fuzzy nature of maps develops a *shadow map of certainty*, assessing map accuracy throughout its spatial extent. Figure 10.1 illustrates such an approach using traditional soil and forest maps (left side) and their corresponding maps of certainty (right side). The shaded features on the base maps indicate soil and forest types 2 and 5 (insets a and c, respectively). Traditionally, we must assume that soil type 2 occurs as one consistent glob in the right portion and forest type 5 occurs as three little globs precisely located as shown—a couple of little white lies.

Just ask the soil or forest photo interpreter who drew the maps. The boundaries are likely somewhere near their delineations, but not necessarily right on. It's their best guess, not the latitudes and departures taken with a surveyor's transit. The challenge for GIS is to differentiate

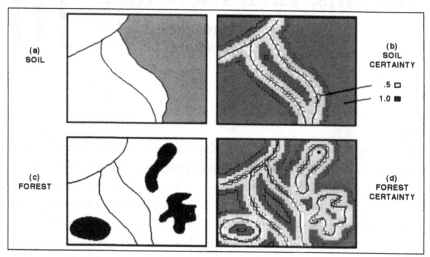

Figure 10.1. Maps of soil and forest cover (left) can be linked to their relative certainty (right). Lighter tones indicate less certain areas around spatial transitions.

that type of mapped data (*interpreted*) from precise map renderings (*measured*), such as surveyed property boundaries. Furthermore, a geographically distributed assessment of certainty should accompany each map—sort of glued to the bottom of the interpreted map's features.

On the right side of the figure, the lines indicate the implied feature boundaries, and the shaded gradient depicts uncertainty as a function of proximity to an implied boundary. The lightest grey areas near the boundaries are assigned a relatively low probability (.5) of correct classification, whereas the interior darker greys are assigned the highest value (1.0). The shades in between represent increasing likelihood (.7 and .9, respectively). This approach reflects a reasonable first-order assumption that areas around soil and forest transitions are the least certain, while feature interiors are the most certain.

In a raster system, the shadow map is generated by "spreading" the boundary locations to a specified distance. The resulting proximity map is renumbered to indicate probability of correct classification—from .5 for boundary locations to 1.0 for locations more than 100 meters away. In that instance, a linear function of increasing probability was used consistently throughout both maps. If warranted, however, a separate function could be developed for each type of boundary transition (e.g., if soils 2 and 3 are easy to delineate, uncertainty might extend only half as far from their implied boundary). A nonlinear function, such as inverse-distance-squared, could be used. Also, if a feature is "pocketed" frequently with other types, the interior certainty value might not attain 1.0 probability.

Admittedly, all that sounds a bit far-fetched, as well as a lot of work for man and machine, but it does illustrate the ability of GIS to map a continuum of certainty, as well as simply feature location—an initial step toward the honest map. Subsequent chapters will look at other means for certainty mapping and its use in error propagation modeling.

1. Monmonier, Mark. *How to Lie with Maps*. The University of Chicago Press, 1991.

For an overview of map certainty and error propagation issues, see "Overlaying Maps and Characterizing Error Propagation," in Berry, *Beyond Mapping: Concepts, Algorithms, and Issues in GIS*, GIS World Books 1993, or "Beyond Mapping," *GIS WORLD*, November-December 1991. Those with tMAP software should review TMAP6.CMD tutorial.

Spawning Uncertainty 11

Characterizing Error Propagation

Chapter 10 developed the concept of a *shadow map of certainty* to express the fuzzy nature of some map boundaries. Now let's extend that concept to *error propagation* when combining maps. That strikes at the bread-and-butter of a GIS—identifying areas of map coincidence. The simple intersection of lines on two maps isn't sufficient, however, for overlaying uncertain maps. Heck, if you weren't sure about either map's boundary placement, why would you be certain about their composite pile of spaghetti? What's needed is a composite shadow map of certainty that tells the user where the joint estimate is likely right on and where it's likely less valid—an honest error assessment.

Figure 11.1 shows the joint coincidence (overlay) of the soil and forest maps described in Chapter 10. The maps on the left identify the "son and daughter polygons" spawned in the conventional overlay process. The process involves the mathematical intersection of the locational tables associated with the two maps. The upper-left map graphically portrays the

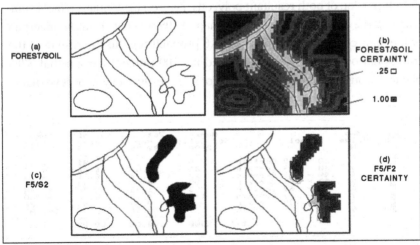

Figure 11.1. The coincidence of the soil and forest maps can be linked to their joint uncertainty (right). Lighter tones indicate less certain areas around the coincidence of feature boundaries.

results, but keep in mind that the GIS "sees" a massive table of X,Y coordinates. The coordinates are grouped into the new polygons, and the attribute tables for the soil and forest maps are merged. The result is a link between each new polygon and its joint soil/forest attribute.

The bottom-left map shows the results (filled polygons) of a geographic search for all locations that are soil type 2 and forest type 5. To achieve this feat, the computer merely searched the composite attribute table for the desired joint condition, then plotted the coordinates of the S2F5 polygons to the screen and filled them with a vibrant color. There, that's it—quick, clean and right on. More importantly, it's comfortable. It's the same thing you'd do if you were armed with transparent sheets, pens, light table and vast amounts of patience.

But how good is the result, considering that uncertainty exists in both of the input maps? Can the GIS account for the propagated error based on the certainty maps? A first-order approximation of the propagated error involves computing the *joint probability* by simply multiplying the soil-certainty map times the forest-certainty map. The right side of figure 11.1 shows the resultant distribution of certainty. The lightest grey tone is the least certain (.5 * .5 = .25) and represents areas of boundary coincidence—not-too-sure and not-too-sure means *really* not-sure. The darker tone identifies areas of relative certainty (1.0 * 1.0 = 1.0).

The bottom-right map isolates the certainty for the geographic search of soil 2 and forest 5. Note that most of the uncertainty is concentrated in the left portion of both resultant polygons. Figure 11.2 contains a table summarizing these data. Nearly half the map (47.06 percent) is fairly uncertain of the joint condition (S2F5 less than .6 probability). But the results of the traditional overlay procedure implied it's 100 percent certain that S2F5 occurs throughout the resultant polygons. Honestly, can you believe that?

Map Color	Map Value	Label		Number of Cells	Pct of Map	Pct of Area	Cumm. Pct
	0			2194	87.76	--	--
	:			--	--	--	--
	2	.20 - .29	Certain	4	.16	1.31	1.31
	3	.30 - .39	Certain	21	0.84	6.86	8.17
	4	.40 - .49	Certain	19	0.76	6.21	14.38
	5	.50 - .59	Certain	100	4.00	32.68	47.06
	6	.60 - .69	Certain	33	1.32	10.78	57.84
	7	.70 - .79	Certain	118	4.72	38.56	96.40
	8	.80 - .89	Certain	2	0.08	.65	97.06
	9	.90 - .99	Certain	9	0.36	2.94	100.00
			Total	2500	100.00	100.00	

Figure 11.2. A table summarizes S2F5 coincidence certainty. Note that nearly half the coincidence map (47.06 percent) is fairly uncertain of the joint condition (S2F5 less than .6 probability).

For additional reading, see "Data Models and Data Quality: Problems and Prospects," M.F. Goodchild, *Environmental Modeling with GIS*, (Oxford University Press, 1993) and "Uncertainty Issues in GIS," session theme (three papers), *GIS '94 Proceedings*, Polaris Conferences, Vancouver, Canada; distributed by GIS World Books, Fort Collins, Colorado.

Dis-Information

*Derivation of Map Certainty
for Interpolated Data*

12

"Dis information ain't right...blow 'em away Bugsy." Sounds like a line from an old gangster movie. But it's more subversive than that. Dis-information is inaccurate data that appear genuine and are used as if they were accurate. In some respects, that describes a large portion of the spatial data populating our GIS databases. Chapters 10 and 11 developed the concept of a *shadow map of certainty* and its use in *error propagation modeling*. The discussion focused on discrete map types (soil and forest) and the probability of correct boundary placement. Now let's turn our attention to continuous surface data and related techniques for assessing error.

Continuous mapped data, such as elevation or barometric pressure, are characterized best as surfaces (vs. the traditional mapping features of points, lines, and polygons). These surfaces are derived primarily through *spatial interpolation*. If you're not too "techno-numbed," review the December 1993-March 1994 series of "Beyond Mapping" columns in *GIS WORLD* discussing spatial interpolation. They establish how the estimates of a mapped variable are derived from field data. But what about the corresponding shadow map of certainty? Where are the estimates good predictors? Where are they bad?

Figure 12.1 identifies the important considerations in developing a map of certainty, based on the same field data used in the earlier interpolation chapters. First, note the circlular window encompassing the 16 field samples of animal activity. All interpolation procedures establish some sort of "roving window" to identify the measurements to be used in the computations. The windows need not be circular, but can take a variety of shapes and even change shape as they move. In that example, the window is large enough and positioned to capture all the data when it's operating in the center of the project area. Locations toward the edges of the map must work with less data (half the circle).

That is a first-order consideration in assessing certainty—the number of sample values used. It's a bit more complex, but it makes common sense that if there's only one field measurement in the window the esti-

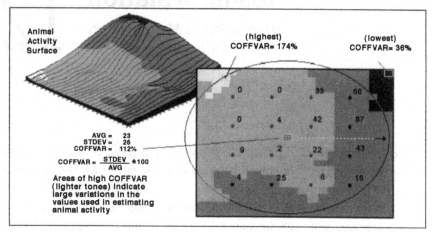

Figure 12.1. Statistics, such as the Coefficient of Variation (COFFVAR), can be used to assess interpolation conditions and generate a map of certainty. The lighter tones indicate areas of less certainty.

mate might be less reliable than if there are several. Another considera-tion is the positioning of the measurements. If the values are at the win-dow's extremities, the estimate might be less reliable than if they were close to the center (location to be interpolated). The "weighted nearest-neighbor" algorithm considers the relative distances to the measure-ments, with the average of their distances providing some insight into the estimate's certainty.

In addition to window shape and data composition, the data values themselves can contribute to certainty assessment. The figure depicts an implied appraisal based on the data's Coefficient of Variation (COFF-VAR), a statistic that tracks the relative variation in the data used in the interpolation. It computes both the average (typical response) and the standard deviation (variability in responses), then computes the per-centage of variability surrounding the average. Interpolated locations based on variable data within the window are assumed to be less reli-able than locations with values that are about the same.

Figure 12.1 identifies four equal steps in variability from 0 percent through 200 percent. The highest COFFVAR is 174 percent (viz., not too sure) at the top left, while the lowest is 36 percent at the top right (viz., more sure). The map was draped on the interpolated surface of animal activity so the viewer can see the number of animals as the height of the surface combined with the colored classes of map certainty. It's interesting that the greatest variability (darker red tones) corresponds to the areas of lower animal activity. Any ideas why?

True, the window in these areas captures the 0's in the northwest corner, as well as some of the high values to the east. Foremost in your reasoning, however, should be the simplicity of that technique. For starters, a "weighted" COFFVAR considering value positioning might help, or a composite statistic considering the size of the window, number of values, their positioning and their variation. How about their alignment with the trend in the data? How might sampling method affect results? What about measurement error vs. procedural error? Whew!

In reality, certainty assessment is a complex area in which spatial statistics scratches the surface. Procedures such as kriging generate a set of shadowed errors each time a set of field measurements is interpolated. Most programs simply discard this valuable information, because the focus is on the estimated surface itself. "Heck, what would users do with a map of error anyway? Let's give them another 256 colors instead."

The important point for mere-mapping-mortals is not an in-depth understanding of statistical theory, but the recognition that maps contain uncertainty and that procedures are being developed to characterize error distribution and propagation. Yep, GIS has launched us beyond mapping as many of us remember it. Strap yourself in, because it's bound to be an exciting, bumpy ride.

Those with the TMAP software should review TMAP1.CMD tutorial on spatial interpolation, and try entering CMD: SCAN DATA WITHIN 13 COFFVAR IGNORING -99 FOR COFFVAR13 to generate figure 12.1.

Empirical Verification

Assessing Mapping Performance Through Error Matrix and Residual Analysis

13

If you want to turn a GIS specialist ashen, suggest taking a map to the field for a little "empirical verification." You know, go to a location and look around to see if the map is correct. If you do that systematically at a lot of locations (keeping track of the number of correct classifications and the nature of the incorrect classifications), you'll learn a lot about the map's certainty. Chapters 10, 11, and 12 discussed ways you could get the computer to guess about certainty—but *empirical verification* uses "ground truth" to directly assess mapping performance.

Consider a typical GIS map, such as soil or forest type. What do you know about its certainty? Usually nothing if you're a typical user and just clicked on a map in a scroll list. It pops up with finely etched features filled with vibrant colors. But try looking beyond the image to its real-world accuracy. Figure 13.1 identifies a couple of ways to do this based on an *error matrix* summarizing the correct and incorrect cells for a set of test plots.

Suppose you had a forest map with categories Pine, Oak, and Fir. After swatting mosquitos and cursing the heat for a couple weeks, you

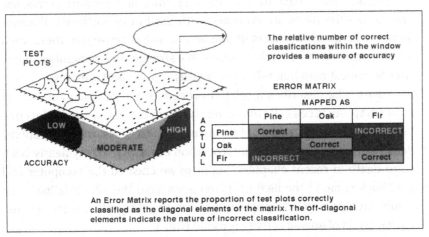

Figure 13.1. An Error Matrix reports the proportion of correct classifications along the diagonal and the nature of the incorrect classifications in the off-diagonal elements.

assemble the field verification data into the matrix shown. In the first cell, record the number of times it was *mapped* as pine when you *actually* stood in pines (PINE-PINE, correct). Record the errors for pines in the next two cells of the column—the number of times the map said you were in pines, but actually you stood in oaks or firs (PINE-OAK and PINE-FIR, incorrect). The correct and incorrect results for the Oak and Fir columns are recorded similarly. Now normalize all of the data in the matrix to the total number of test plots for each category so it is expressed in percent.

Note that the proportion of correct classifications are along the diagonal, and the nature of the incorrect classifications are in the off-diagonal elements—a good summary of overall mapping performance. The off-diagonal information on errors is particularly useful.

But what about the distribution of the errors throughout the project area? A first-order guess involves moving a summary window around a map of the results of the test plots. If you find that the preponderance of the mistakes were in the northwest, you might consider additional field checking in that area. Or, possibly that area was mapped by an individual needing a refresher course (or a pink slip). A useful modification to the procedure is to note just the errors in the roving window that involve the category at the window's focus. That gives you an idea of how well that classification is doing in its general vicinity. It may be that pines frequently are misclassified in the northeast, but exhibit good classification in the southwest.

You also could have the window keep track of the nature of misclassifications—the pines are confused with firs in the northeast. Because there are so many fir stands in that area it makes sense that there are a lot of errors. All that might sound a bit strange, but remember, we're after an honest map that tells us more than just its best guesses.

Figure 13.2 describes a related procedure for verifying continuous data (map surfaces). *Residual analysis* investigates the difference between a predicted value and the actual value for a set of test locations. Recall the example of spatial interpolation of animal activity nearly beaten to death in recent chapters. Suppose we cheated the computer and held back some of the field measurements from the interpolation. That would give us a good opportunity to verify its performance against known levels of animal activity (ground truth).

Figure 13.2 identifies 13 ground truth locations of measured animal activity. The difference between the predicted and actual values at a

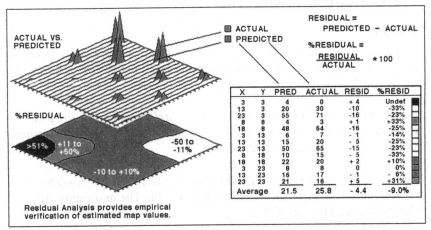

Figure 13.2. Residual Analysis investigates the difference between a predicted value and the actual value for a set of test locations.

location identifies the *residual*. In the figure, the sample plot at X= 23 and Y= 13 is identified in the 3-D map. Its residual is computed as −15 (i.e., 50–65= −15). The sign of the residual indicates whether the guess was too high (+) or too low (−). The magnitude of the residual indicates how far off the guess was. The percent residual merely normalizes the magnitude of error to the actual value; the higher the percentage, the worse the performance.

To generate the %RESIDUAL map (bottom left) the %RESID values in the table were, in turn, interpolated for a map of the percent residual. Note that the northeast and southwest portion of the project area seemed to be on target (+10 percent error), whereas the southeast and northwest portions appear less accurate. The extreme northwest portion is way off (>51 percent error).

I wonder why? Like Chapter 12, the answer probably lies in the assumptions and simplicity of the analysis procedure as much as it lies in the data themselves. Spatial statistics is a developing field for which theory and practical foundation has yet to be set in concrete. At the moment we may have the cart (GIS) in front of the horse (science), but the idea of an honest map boldly displaying its errors will become a reality in the not too distant future...mark my words.

Recommended Reading

Books

Berry, J.K. "Overlaying Maps and Characterizing Error Propagation." Topic 6 in *Beyond Mapping: Concepts, Algorithms, and Issues in GIS*, Fort Collins, CO: GIS World Books, 1993.

Burrough, P.A. "Data Quality, Errors, and Natural Variation." Chapt. 6 in *Principles of Geographical Information Systems for Land Resources Assessment*, Oxford, UK: Clarendon Press, 1986.

Burrough, P.A. "Accuracy and Error in GIS." In *The AGI Source Book for Geographic Information Systems*, 87-91. London, UK: The Association for Geographic Information, 1994.

Chrisman, N.R. "The Error Component in Spatial Data," in *Geographical Information Systems: Principles and Applications*, eds. D.J. Maguire, M.F. Goodchild, and D.W. Rhind, Vol 1, 165-74. Essex, UK: Longman, 1991.

Goodchild, M.F. "Data Models and Data Quality." In *Environmental Modeling with GIS*, eds. Goodchild, Parks, and Steyaert, 94-103. Oxford, UK: Oxford University Press, 1993.

Goodchild, M.F. "Issues of Quality and Uncertainty," in *Advances in Cartography,* ed. Muller, J., 113-139. New York, NY: Elsevier, 1991.

Lodwick, W.A. "Developing Confidence Limits on Errors of Suitability Analyses in Geographical Information Systems," in *Accuracy of Spatial Databases*, eds. Goodchild, M.F., and S. Gopal, 69-78. New York, NY: Taylor and Francis, 1989.

Journal Articles

Altman, D. "Fuzzy Set Theoretic Approaches for Handling Imprecision in Spatial Analysis." *International Journal of Geographical Information Systems* 8: 271-89 (1994).

Dunn, R., A.R. Harrison, and J.C. White. "Positional Accuracy and Measurement Error in Digital Databases of Landuse: An Empirical Study." *International Journal of Geographical Information Systems* 4: 385-398 (1990).

Goodchild, M.F., S. Guoqing, and Y. Shiren. "Development and Test of an Error Model for Categorical Data." *International Journal of Geographical Information Systems* 6(2): 87-104 (1992).

Heuvelink, G.B.M., and P.A. Burrough. "Error Propagation in Cartographic Modeling using Boolean Logic and Continuous Classification." *International Journal of Geographical Information Systems* 7: 231-246 (1993).

A FRAMEWORK FOR GIS MODELING

ESSENTIAL CONCEPTS AND

PRACTICAL EXPRESSIONS

We must be able to digest the
mass before it becomes a mess.
—*John Griffiths*

The use of GIS to model complex spatial relationships is rapidly increasing. Our understanding of the types and approaches of models, however, is based on decades of nonspatial modeling experience. This section presents a classification framework for GIS models and a flowcharting methodology.

What's in a Model?

Types and Characteristics of GIS Models

14

I conduct a lot of GIS courses and workshops. As you might imagine, they frequently move beyond the fundamental concepts to futuristic musings. One topic consistently captures the imagination of participants and dominates informal discussion (you know, the elevated B.S. in the sunken lounge): What are the types and characteristics of GIS models? The accompanying outline is the current state of a "sourdough" handout used to provoke this impassioned discussion. Keep the following questions in mind while you review the outline:

- Do you know of any model types or characteristics missing from the outline? Are any in the outline misrepresented?

- The following terms also are used to describe models: physical, atomistic, holistic, constrained, fragmented, dispersed, data, analytical, diffusion, scale, optimizing, simulation, analytical, process, synthetic, systems, flow, statistical, mathematical, hierarchical, binary... Can you explain what's meant by these terms? Are any relevant? Where might they fit into the outline?

- Do you see any utility in developing a comprehensive classification scheme for GIS modeling, or is this just another esoteric and academic (gee, that might be redundant) exercise? Who would benefit from such an outline?

Types and Characteristics of GIS Models

1. Modeling

A model is a representation of reality in either *material* form (tangible representation) or *symbolic* form (abstract representation); GIS modeling involves symbolic representation of *locational* properties (WHERE), as well as *thematic* (WHAT) and *temporal* (WHEN) attributes describing characteristics and conditions of space and time.

2. General Types of Models: *Structural and Relational*

Structural. Focuses on the composition and construction of things; *object* and *action*.

Object. *Static entity-based*, forming a visual representation of an item, e.g., an architect's blueprint of a building. Characteristics include scaled, two- or three-dimensional, symbolic representation.

Action. *Dynamic movement-based*, tracking space/time relationships of items, e.g., a model train along its track. Characteristics include time-slices, change detection, transition statistics, and animation.

Relational. Focuses on the interdependence and relationships among factors; *functional* and *conceptual*.

Functional. *Input/output-based*, tracking relationships among variables, e.g., storm runoff prediction. Characteristics include cause/effect linkages, hard science, and sensitivity analysis.

Conceptual. *Perception-based*, incorporating fact interpretation and value weights, e.g., suitability for outdoor recreation. Characteristics include heuristics (expert rules), soft science, and scenarios.

3. Types of GIS Models: *Cartographic and Spatial*

Cartographic. Automation of *manual techniques* that traditionally use drafting aids and transparent overlays, e.g., a map identifying locations of productive soils and gentle slopes using binary logic expressed as a geo-query.

Spatial. Expression of *mathematical relationships* among mapped variables, e.g., a map of surface heating based on ambient temperature and solar irradiance using multivalue logic expressed as variables, parameters, and relationships.

4. GIS Model Characteristics: *Scale, Extent, Purpose, Approach, Technique, Association, and Aggregation*

Scale. *Micro* and *Macro*

Micro. Contains high resolution of space, time, and/or variable considerations governing system response, e.g., a 1:1,000 map of a farm with the crop specified for each individual field revised each year.

Macro. Contains low resolution of space, time, and/or variable considerations governing system response, e.g., a 1:1,000,000 map of land use with a single category for agriculture revised every 10 years.

Extent. *Complete* and *Partial*

Complete. Includes entire set of space, time, and/or variable considerations governing system response, e.g., a map of an entire watershed or river basin.

Partial. Includes subsets of space, time, and/or variable considerations governing system response, e.g., a standard topographic sheet with its artificial boundary capturing limited portions of several adjoining watersheds.

Purpose. *Descriptive* and *Prescriptive*

Descriptive. Characterization of the direct interactions of system components to gain insight into system processes (understand), e.g., a wildlife population dynamics map generated by simulation of life/death processes.

Prescriptive. Characterization of direct and indirect factors related to system response used in determining appropriate management action (decide), e.g., a campground suitability map based on interpretation of landscape features.

Approach. *Empirical* and *Theoretical*

Empirical. Based on reduction (*analysis*) of field-collected measurements, e.g., a map of soil loss for each watershed in a region generated by spatially evaluating the Universal Soil Loss Equation.

Theoretical. Based on the linkage (*synthesis*) of proven or postulated relationships among variables, e.g., a map of spotted owl habitat based on accepted theories of owl preferences.

Technique. *Deterministic* and *Stochastic*

Deterministic. Direct evaluation of a defined relationship (results in a single repeatable solution), e.g., a wildlife population map based on one model execution using a single "best" estimate to characterize each variable.

Stochastic. Simulation of a probabilistic relationship (results in a range of possible solutions), e.g., a wildlife population map based on the average of a series of model executions using probability functions to characterize each variable.

Association. *Lumped* and *Linked*

Lumped. The state/condition of each individual location is *independent of other map locations* (point-by-point).

Linked. The state/condition of an individual location is *dependent on other map locations* (vicinity, neighborhood, or region).

Aggregation. *Cohort* and *Disaggregated*

Cohort. Executed for *groups of objects* having similar characteristics, e.g., a timber growth map for each management parcel based on a look-up table of growth for each specific set of landscape conditions.

Disaggregated. Executed for each *individual object*, e.g., a map of predicted biomass based on spatially evaluating a regression equation in which each input map identifies an independent variable, each location a case and each value a measurement (usually raster-based grid cells).

Temporal. *Static* and *Dynamic*

Static. Treats time as constant and model variables *do not vary over time*, e.g., a map of timber value based on forest inventory and relative access to existing roads.

Dynamic. Treats time as variable and model variables *change as a function of time*, e.g., a map of the spread of pollution from a point source.

In Chapter 15, we'll translate the outline into a generalized Classification Guide for GIS Models. Sound like fun, or more pedagogical pomposity?

The GIS Modeling Babble Ground 15

A Classification Guide for GIS Models

As you may recall from Chapter 14, there are many dimensions to GIS modeling. Modeling is as personal as the underwear you buy or the politics you support. GIS modeling perspectives are the result of the data you keep and the things you do. A county clerk, city engineer, forester, and market forecaster work with radically different data for many diverse purposes. In the applied arena, GIS modeling means different things to different people—hence the "babble-ground" lines are drawn in the sand of confusion.

If you strip away the details of specific applications, however, common threads appear among the GIS models themselves and the modeling processes undertaken. Chapter 14 attempted to capture some of the more important threads. The factors discussed are stripped of their verbiage and summarized in figure 15.1.

```
                              MODEL
                          (Representation)

                    MATERIAL .......... SYMBOLIC
                     (Tangible)          (Abstract)

General Model
TYPE:          STRUCTURAL                RELATIONAL
               Object. . . . . . . . . Action    Functional . . . . . . . . . Conceptual

GIS Model
TYPE:                                    Cartographic . . . . . . . . . Spatial

GIS Model
CHARACTERISTICS:

   SCALE                              Micro . . . . . . . . . Macro
   EXTENT                           Complete . . . . . . . . . Partial
   PURPOSE                        Descriptive . . . . . . . . . Prescriptive
   APPROACH                         Empirical . . . . . . . . . Theoretical
   TECHNIQUE                    Deterministic . . . . . . . . . Stochastic
   ASSOCIATION                       Lumped . . . . . . . . . Linked
   AGGREGATION                       Cohort . . . . . . . . . Disaggregated
   TEMPORAL                          Static . . . . . . . . . Dynamic
```

Figure 15.1. Classification guide for GIS models.

One of the most frustrating aspects of any classification scheme is being forced to assign something to one of two choices (binary logic). It's like those dumb questions on the aptitude tests—not everything is black and white. In the classification guide, the descriptors for each factor identify opposing extremes. The dots separating the extremes provide a range of possible responses—you simply place an "X" at the appropriate spot along the continuum. The dichotomies have been arranged so a clustering of marks toward the left indicates models that are comprehensible without a Ph.D. in complex studies.

Let's tackle an easy example and force responses to the extremes. Consider the famous sculpture of Venus de Milo. Sure it's a model (abstraction), or she sure has us all fooled by sitting so still. Within the limits of the classification guide, she falls into the following categories:

- *Material* (one big piece of marble; no abstract symbols here)

- *Structural* (model characterizes her construction; don't know about her relationships)

- *Object* (visual rendering of just her; no movable parts)

Granted, she's not a GIS model. If she was, however, she could be categorized as follows:

- *Cartographic* (manual techniques; no wimpy mathematics)

- *Micro* (about a 1:1 scale; unless she's a scaled version of Goliath's mom)

- *Partial* (missing arms and legs; or maybe they were nicked in a move)

- *Descriptive* (wow, and how; doesn't tell you what to do…she's just a rock)

- *Empirical* (direct measurement; or the sculptor had an active imagination)

- *Deterministic* (direct, single solution; hips and shoulders have no chance of being attached elsewhere)

- *Linked* (the hip bone is connected to the thigh bone…; can't talk about her chin without noticing her eyes)

- *Disaggregated* (one-of-a-kind; though millions strive for a favorable comparison)

- *Static* (hasn't changed for centuries; the whole effect is dynamite, but not dynamic)

Now let's try a tougher one: an animated set of maps predicting wildfire growth for hourly time steps. figure 15.2 indicates "refined" response positioning along each of the scales, whereas the following discussion identifies the extremes. The first part is easy, as the fire model leans toward the following categories:

```
                            MODEL
                        (Representation)

                MATERIAL  . . . . . . . . .☒ ABSTRACT
                 (Tangible)                 (Symbolic)

General Model TYPE:
        STRUCTURAL                    RELATIONAL
   Object . . . . . . . . . Action    Functional .☒. . . . . . . Conceptual

   GIS Model TYPE:
                                      Cartographic  . . . . . .☒. . Spatial

   GIS Model CHARACTERISTICS:

      SCALE                      Micro . .☒. . . . . . Macro
      EXTENT                  Complete . . . . . . .☒. Partial
      PURPOSE              Descriptive . .☒. . . . . . Prescriptive
      APPROACH              Empirical .☒. . . . . . . Theoretical
      TECHNIQUE         Deterministic ☒. . . . . . . . Stochastic
      ASSOCIATION          Distributed . . . . . . .☒. Linked
      AGGREGATION             Cohort . . . . . . .☒. Disaggregated
      TEMPORAL                  Static . . . . . .☒. . Dynamic
```

Figure 15.2. A completed classification guide evaluates an animated set of maps predicting wildfire growth for hourly time steps.

- *Abstract* (or you'd better get a hose)

- *Relational* (depends on several mappable factors, including terrain, vegetation type/condition and weather)

- *Functional* (mostly uses fire science research tracking the relationships among variables)

The more perplexing part involves the following GIS model type and characteristics:

- *Spatial* (a lot of math behind this one)

- *Micro* (considers only the fire front and its immediate surroundings)

- *Partial* (until the fire is extinguished)

- *Descriptive* (unabated fire propagation without fire management actions)

- *Empirical* (based on field-calibrated equations)

- *Deterministic* (based on a defined set of input parameters)

- *Linked* (adjacent parcels next)

- *Disaggregated* (independently considers each burning location and its propagation options)

- *Dynamic* (diurnal and ongoing fire behavior conditions change model variables)

Whew! Now try your hand at "classifying" the following representations of reality and/or your own favorite GIS models:

- Mount Rushmore's faces of the presidents

- A landscape architect's cardboard model of a national park

- An elk habitat map

- A set of seasonal maps of elk habitat

- An elk population dynamics model responding to landscape conditions and predator/prey interactions

- A GIS implementation of the Universal Soil Loss Equation for a watershed

- A GIS implementation of the Horton Overland Flow Equations evaluating surface water runoff for a set of watersheds

- A crop yield prediction map

- Maps of wildfire risk generated each morning

- A dynamic wildfire growth model responding to temperature fluctuations, complex wind vectors and fire abatement actions

Enjoy!

A classic reference for modeling is *Mathematical Modeling with Computers*, by Jacoby and Kowalik, Prentice-Hall, 1980. Ample "poetic license" was used in extending the basic modeling framework to the unique conditions and approaches used in GIS modeling.

Layers to Tapestry

16

Diagramming GIS Logic and Processing Flows

Most people agree there are three essential elements to GIS: data, operations, and applications. To use the technology you need a bunch of digital maps, an analytic "engine" to process the maps, and interesting problems to solve. However, there are different views regarding the relative importance of the three elements. Some people have a *data-centric* perspective, as they prepare individual data layers and/or assemble the comprehensive databases GIS needs. Other people are *operations-centric*, because they lock in on refining and expanding the GIS toolbox of processing and display capabilities. Finally, the *applications-centrics* see the portentous details of data and operations as mere impediments to problem solving. Such is the fractious fraternity of GIS.

In the early years, the data and operations orientations dominated the developing field. As GIS matured, the focus shifted to applications. As a result, more attention is directed toward the assumptions and linkages embedded in our GIS models—the map analysis solutions to pressing problems. In essence, we're weaving our data layers into complex, logical tapestries of map interrelationships. A crucial component of this evolution is an effective mechanism to communicate model logic, as well as processing flow.

Programmers and system analysts routinely use diagramming techniques for communicating data/processing flow. Various approaches include structure and flow charts, as well as data flow, entity relation, control flow, and state transition diagrams. Each technique invokes a subtly different perspective in communicating structure and logic. For example, a data flow diagram emphasizes the processing steps used in converting one data set into another (figure 16.1). The technique uses large circles to symbolize operations, with the lines connecting them representing data sets. Its design draws one's attention to the processing steps over the data states, thereby best serving an operations-centric orientation.

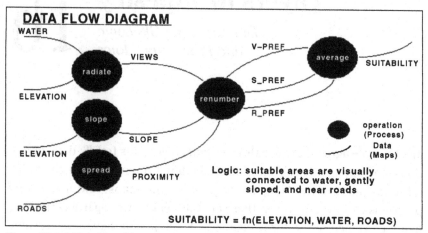

Figure 16.1. A Data Flow Diagram.

Processing-oriented diagrams work well for nonspatial information processing. They relate data about entities through indexed files. In these instances, the specifications in a database query are paramount. Instances of geo-query, such as "Where are all the locations that have slopes greater than 13 percent AND unstable soils AND are devoid of vegetation?" use standard database management systems technology. In such instances, standard diagramming techniques are most appropriate.

However, spatial analysis techniques go beyond repackaging existing data. For example, it's a different story if you want to establish variable-width buffers around salmon spawning streams. You need simultaneously to consider intervening slopes, ground cover, and soil stability as you "measure" distance. If you want to establish a map of visual exposure density to roads, you need to consider maps of the road network and relative elevations at a minimum. These, and myriad other spatial analysis procedures, have strong data dependency. They aren't just setting a few parameters for traditional, nonspatial processing techniques. Spatial analysis is a new kettle of fish. It's dependent on the unique geographic patterns of the data sets involved—definitely data-centric conditions.

A GIS model flowchart, or "map model," takes such a perspective. The top of figure 16.2 uses a flowchart to track the same data/processing steps as shown in the data flow diagram. Maps (i.e., data sets) are depicted as boxes, and operations (i.e., processing steps) are depicted as arrows. Obviously, this focus is data-centric because it draws your attention to the mapped variables. Arguably, it's also applications-centric. Most GIS users have experience with manual map analysis techniques. They've struggled laboriously with rulers, dot grids, and transparent overlays to draft new maps that better address a question at hand. For example, you may have circled areas where the elevation contour lines are close together to create a map of steep slopes. In doing so, attention is focused on the elevation data and the resultant circles inscribed on the transparent overlay—the input and output maps.

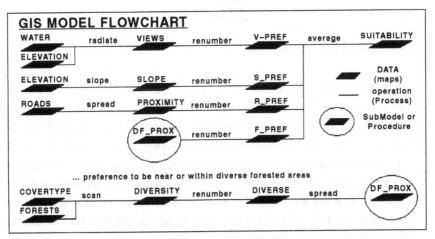

Figure 16.2. A GIS Model Flowchart.

The bottom portion of figure 16.2 shows a logic modification incorporating a preference to be near or within diverse forested areas. A neighborhood operation (scan) assigns the number of different vegetation types (COVERTYPE) within the vicinity of each forested location (FOREST). Areas of high diversity are isolated (renumber), and a proximity map from these areas (DF_PROX) is generated for the entire project area. Because several models might share this command set, it's stored as a generalized procedure and is attached using the *SubModel or Procedure* flowcharting "widget."

Figure 16.3 identifies a processing modification to the model. In this example, a display of the SUITABILITY map with road vectors (ROAD.BLN) graphically overlaid is used as a backdrop for the user to manually draw a potential set of SUITABLE sites. Statistics on the sites (STATS.TBL) are presented, and the user can either accept them or redraw another set of potential sites. When accepted, the raster map is converted to vectors and stored. The example uses an extended set of *Connector, File, Manual Operation, Conditional Branch* and *Nonspatial Operation* widgets.

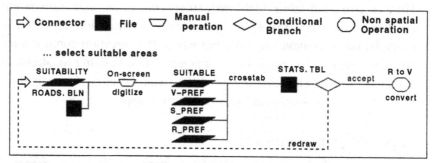

Figure 16.3. Additional flowchart widgets.

So what? All that seems to be "much ado about nothing." It's just a bunch of globs, lines, and silly symbols. Actually, it may be the way for GIS to get out of the black box and into the light of creative applications. General users need a simple flowchart of model logic to understand and appropriately apply a model. A more complex flowchart extending to processing flow is needed by the GIS specialist who wrestles with the actual code. What we all need is a single diagramming technique that can operate at both levels—a simple logical expression that can be embellished with processing flow details.

Recommended Reading

Books

Burrough, P.A. "Methods of Data Analysis and Spatial Modeling." Chapt. 5 in *Principles of Geographical Information Systems for Land Resources Assessment*, Oxford, UK: Clarendon Press, 1986.

Coppock, J.T., and D.W. Rhind. "The History of GIS," in *Geographical Information Systems: Principles and Applications*, ed. D.J. Maguire, M.F. Goodchild, and D.W. Rhind, Vol 1, 21-43. Essex, UK: Longman, 1991.

Densham, P.J. "Spatial Decision Support Systems," in *Geographical Information Systems: Principles and Applications*, ed. D.J. Maguire, M.F. Goodchild, and D.W. Rhind, Vol 1, 403-12. Essex, UK: Longman, 1991.

Tomlin, C.D. "Cartographic Modeling," in *Geographical Information Systems: Principles and Applications*, ed. D.J. Maguire, M.F. Goodchild, and D.W. Rhind, Vol 1, 361-74. Essex, UK: Longman, 1991.

Journal Articles

Berry, J.K. "A Mathematical Structure for Analyzing Maps." *Journal of Environmental Management* 11(3): 317-25 (1987).

Bracken, I., and C. Webster. "Towards a Typology of Geographical Information Systems." *International Journal of Geographical Information Systems* 3(2): 137-152 (1989).

Burrough, P.A. "Matching Spatial Databases and Quantitative Models in Land Resource Assessment." *Soil Use and Management* 5: 3-8 (1989).

Burrough, P.A. "Development of Intelligent Geographical Information Systems." *International Journal of Geographical Information Systems* 6(1): 1-11 (1992).

Weibel, R., and B.P. Buttenfield. "Improvement of GIS Graphics for Analysis and Decision-Making." *International Journal of Geographical Information Systems* 6(3): 223-45 (1992).

ALTERNATIVE DATA STRUCTURES
OPTIONS BEYOND
RASTER AND VECTOR

Computers aren't people ... they read bar
codes and numbers with fractional parts,
and do floating point arithmetic.
— *Nicholas Negroponte*

At the heart of GIS is its data. How these data are structured, in large part, determines a system's performance, capabilities, and breadth of applications. This section describes alternative approaches to traditional raster and vector data structures.

Are You a GIS Dead Head?

Structuring Traditional Raster Data

17

Even if you're new to GIS, you may have encountered the scholarly skirmishes between the raster heads and the vector heads. Like other religious crusades, the principles in these debates often are lost to mind-sets reflecting cultural exposure and past experience. More often than not, however, most of us just become catatonic when the discussion turns to GIS data structures. But what the heck—it's worth another try.

Let's review the basic tenets of vector and raster data, then extend this knowledge to the actual data structures involved. Vector data use sets of X,Y coordinates to locate three basic types of landscape features: points, lines, and areas. For example, a typical water map identifies a spring as a dot (one X,Y coordinate pair), a stream as a squiggle (a set of connected X,Y coordinates), and a lake as a glob (a set of connected X,Y coordinates closing on itself and implying its interior). Raster data use an imaginary grid of cells to represent the landscape. Point features are stored as individual column/row entries in the grid; lines are identified as a set of connected cells; and areas are distinguished as all of the cells comprising a feature.

That traditional representation constrains geographical phenomena to three user-defined conditions (points, lines, and areas) and two GIS expressions (vector and raster). I bet this conceptual organization is fairly comfortable, and might even be familiar. But that's only half the problem—the user/GIS representation has to be translated into a database/hardware structure. At that step most of us simply glaze over and leave such details to the GIS jocks. Actually, the concepts aren't too difficult and they can help us understand much about different systems, the frustrations we may encounter, and the future direction of GIS.

Let's consider some structures for raster data. The left side of figure 17.1 shows an imaginary grid superimposed on a typical soil map (more appropriately termed a "data layer"). The center portion of the figure identifies a matrix of numbers with a numerical value assigned to each cell. In that case, the value represents a particular type of soil, and its position in

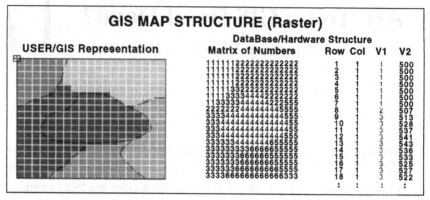

Figure 17.1. A file containing a matrix of numbers (attribute values) characterizes each cell of an imaginary grid. An alternative structure uses a standard database file containing the column/row identifiers for each cell followed by attribute fields, such as soil type and elevation.

the matrix indicates its location. To the computer, however, the matrix isn't a two-dimensional array but simply one long list of numbers. The first number represents the upper-left corner of the matrix and the rest are ordered from left to right, top to bottom. Another map for the same area, say elevation, would be stored as a separate ordered sequential file: This is the simplest and most frequently used raster data structure.

An offshoot of raster structure is used for most remote sensing data. Each cell in the grid represents a small area of Earth's surface where a satellite collected spectral data. The numbers record relative amounts of energy, such as blue, green, and red light radiating from the surface. The set of values for each energy level represents a single data layer that can be arranged as a matrix and stored separately. These data, however, are stored more efficiently as an interlaced matrix, with all of the measurements for each cell stored sequentially. For example, the first three values might represent the blue, green, and red light measurements for the upper-left cell, with the following triplets of values for the other cells sequenced as before—left to right, top to bottom.

The interlaced structure has a significant advantage in point-by-point processing, because all information is contained in a single file and available as the computer methodically steps through the matrix. The computer doesn't have to open three separate files, then read blue, green, and red values scattered all over the disk. As a result, you have a lot less disk-thrashing and a happier computer. That may not seem like a big deal, but considering that a typical Landsat Thematic Mapper scene contains seven data layers for about 36 million cells (that's more than 250 million numbers), even a slight increase in storage efficiency is cyber heaven.

The interlaced structure might be neat and tidy for remote sensing data, but it's inappropriate for a general GIS. First, it's tough to add a new map. It means extra room must be made to insert the new values by reading the first three values from the original file, writing them to a new file, inserting the first new value, and then repeating the process for the other million or so cells. Oh yes, then delete the original file. And you have the same problem if you want to delete a map. Second, because the information for each data layer is dispersed (every third value), it's difficult to compress the redundancy found in a typical map. Finally, any processing involving neighboring cells requires extra work as the computer must continually jump back and forth in the file to get values for the cells above, below, right, and left of the target cell. In short, the interlaced structure is best for specialized applications involving a fixed number of maps constrained to point-by-point processing.

So what else do we have in raster structures? Consider the right side of figure 17.1. The structure uses a standard database file (a "database table") with the column/row entries of the matrix explicitly stored as "fields" (i.e., separate columns). The subsequent fields contain the listing of values for various data layers. Note that the set of soil values under V1 corresponds to the left-hand column of the matrix. If there was room in the figure to list the rest of the values in the field, the next set would replicate the next column to the right in the matrix, then the next, and so on. The V2 listing depicts similarly organized elevation values.

Now comes the advantage. Suppose you want to find all locations (i.e., cells) that contain soil type 4 and are more than 550 feet in elevation. Simply enter a Structured Query Language (SQL) command and the computer searches V1 and V2 for the specified condition. A new field (V3) will be appended containing the results. It's easy, because you use a standard database file under the control of a standard database management program. A standardized structure makes it easy for GIS programmers, because they don't have to write all the code that's already in the database "engine." Also, it allows you to store and process text string designations, as well as numerical values. More importantly, it makes it easy on the user, because the command uses the same format as a normal office database.

The problem lies with the computer. It hates appending new fields to an existing table. Also, the number of fields in a single table is constrained. The solution is a series of indexed tables, with each cell's designation serving as the common link. With an indexed structure the computer can easily "thread" from one table to another. Actually, there are good arguments for storing each data layer as a separate indexed table. Creating, modifying, or deleting a map is a breeze, because it affects only one table rather than a field embedded in a complex table. That seems to brings us back to where we began—one map, one file. In that instance however, each map is a standard indexed database table with all of the rights, privileges, and responsibilities of your office database. It puts raster GIS where it should be—right in the middle of standard database technology. As we'll see in Chapter 18, vector GIS has been there all along.

Raster is Faster, but Vector is Correcter

Structuring Traditional Vector Data

18

Your computer really loves raster data—a cell on one map is at the same position on all others. A couple of "hits-to-disk" and it knows everything about a cell location. A few more hits up, down, left and right and it knows everything about a location's entire neighborhood. In fact, it can "walk" from one location to another and find everything it needs to know along the way, right from the hard disk. Its world is pre-defined in little byte-size pieces that are just right.

However, the computer-endearing qualities of consistency and uniformity put raster data at odds with the human psyche (and a lot of reality). We see the unique character of each map feature—a cute little jog here, a little bulge there. The thought of generalizing these details into a set of uniform globs is cartographic heresy.

So what does it cost your computer, in terms of data structure, to retain the spatial precision you demand? First, because every map feature is unique, a complex data structure is required. Consistency and uniformity are out; uniqueness and irregularity are in. More importantly, processing involves threading through a series of linked files (called "tables" in database-speak), mathematically constructing map features, calculating the implied coincidence, then reconstructing the new data structure linkages; all that just to know the property line isn't 100 feet over there ... picky, picky.

Figure 18.1 identifies the basic elements of vector data structure. It begins with a points table, attaching coordinates to each point used in the construction of map features. Most systems use latitude and longitude as their base coordinates. That's a good choice, as it's a spherical coordinate system that accurately locates points anywhere on Earth's surface. It's a problem, however, when you want to relate points, such as measuring distances, bearings, or areas.

In three dimensions, seemingly simple calculations involve solid geometry and ugly equations that bring even powerful computers to their knees. Plus, the three-dimensional answers can't be drawn on a flat

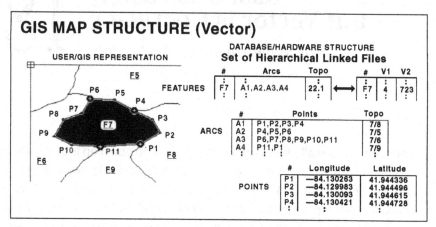

Figure 18.1. A set of files links coordinates, arcs, and features to describe location. A standard database file links each feature to its attributes, such as soil type and elevation.

screen. The solution is to carry a user-specified map projection scheme and planar coordinate system (e.g., Universal Transverse Mercator), then translate on-the-fly. Now the computer can work in any two-dimensional rendering you choose and easily display the results on your screen or plotter ... happy computer, happy you.

For point features, the point table and its two-dimensional translation specifications are linked directly to another indexed file containing descriptive information, called "attributes," about each point. If this information depicted soil samples, you could query the attribute table for all of the samples that have a pH less than 7 and available phosphorus exceeding 30 parts per million. The results from the attribute query simply "threads" to the coordinates of the subset of points meeting the conditions, then plots them at blinding speed in the vibrant color of your choice.

Line and area features are a bit more complicated because the various connections among sets of points need to be specified. When you view a human-compatible map of water features you intuitively note which stream is connected to which stream by the network of blue squiggles. You note that lakes are blue globs with a squiggle in and another out. But the computer's point file is just a huge pile of unrelated numbers. The first level of organization is a linked arcs table. That file groups the points into connected sets of arcs, forming the map features. In figure 18.1, points P1, P2, P3, and P4 are connected to form arc A1. Similarly, arcs A2, A3, and A4 are defined by their linked coordinates.

The features table puts it all together in geographic space by linking the arcs to actual map features. In the example, feature F7 is formed by linking arcs A1, A2, A3, and A4. The corresponding arcs table identifies which points are involved, with the coordinates in the points table ultimately tying everything to the ground. At the top of this scheme is a linked info table with the attribute data for each map feature. In the example, feature F7 is identified as having soil type 4 (V1) and an average elevation of 723 (V2).

There, that's not too bad—conceptually. The tough part comes when you try to put it all into practice with about 100,000 polygons. That's when each vendor's "secret ingredients" of the general vector recipe take hold. Without giving away any corporate secrets, let's take a look at some of the "tweaking" possibilities.

In addition to the link to the points, the arc and feature tables often contain topological and other information. For example, note that arc A1 forms a shared boundary between features F7 and F8 as listed in the "Topo" field of the arcs table (7/8). For maps composed of contiguous polygons (e.g., soils, cover type, ownership, and census tracts) a search of this field immediately identifies the adjoining neighbors for any map feature. Many systems store frequently used geometric measurements— such areas as depicted in the "Topo" field of the features table (22.1 acres). The alternative to these tweaks in data-structure design is a lot of computational thrashing and bashing each time they're needed.

Line networks use topological information to establish which arcs are interconnected and the nature of their connections. In a stream network it depicts the direction of water flow. In a road network it characterizes all possible routes from any location to all other locations. To describe the linkage embedded fully, however, a new element must be introduced: the node. These special points are indicated in the figure as the large dots at the ends of each arc (P1, P4, P6, and P11). Nodes represent locations where things are changing, such as the separation of adjacent soil units along a soil boundary. Some systems store nodes in a separate table, while others simply give them special recognition in the points table. The information associated with a node reflects the type of data and the intended processing.

If the length of each arc is stored, the computer can find the distance from a location to all other locations by simply summing the intervening arcs along a route. If an average speed is stored for each arc, the answer will be in travel-time. But what about one-way streets and the relative difficulty of left and right turns at each intersection or node? Attach that information to the nodes and the computer will make the appropriate corrections as it encounters the intersections along a route. Similarly, an accumulated distance from a location to its surroundings can be determined by keeping a running sum of the arc distances, respecting the "turntable" information at each node. Once that's known it's an easy matter to determine the optimal path (shortest time or distance) from any location to the starting point.

All is for naught, though, if your data structure hasn't been tweaked to carry the extra topological and calibration information. It should be apparent that, unlike raster data structures, vector data structures can be radically different. Ingenuity and programming dexterity are critical factors, as is the matching of data design to intended applications and hardware. That's the tough part; there isn't a universal truth in vector data structure. The onus is on you to pick the right one for your applications, then understand it enough to take it to its limits.

19

How Are Your Quads and TINs?

Quadtrees and Triangulated Irregular Networks

The original raster and vector data structures have been around a long time. The basic concepts of representing a landscape as a set of grid cells or a set of connected points are about as old as cartography itself. The technical refinements required for a functioning GIS, however, evolve continually. About the time I think we've reached the pinnacle of data structure design, someone comes out with a new offshoot. The only folks who think they have it all are overzealous marketers. The technical types keep their heads down, constantly looking for more effective ways to characterize mapped data.

Most raster systems can perform run-length compression, which compacts along the rows or the columns. For example, consider the following matrix and its row-compressed translation.

Full Matrix	Run-Length (Row)
111111122222222223	1,7,2,17,3,18
111111122222222233	1,7,2,16,3,18
111111122222222333	1,7,2,15,3,18
111111222222223333	1,6,2,14,3,18
111113333333333333	1,5,3,18
111113333333333333	1,5,3,18
111113333333333333	1,5,3,18
111333333333333333	1,3,3,18
111333333333333333	1,3,3,18

The run-length data structure uses just 44 numbers to represent the 162 numbers in the full matrix. It uses "value through column" pairs of numbers to compact redundancy along a row. For example, the first row is read "value from column 1 (assumed) through column 7, value 2 from column 8 (last column plus one) through column 17, and value 3 from column 18 through column 18." That format is particularly useful in map display. Instead of "hitting disk" for the color-fill-pattern value at each cell, the computer reads the pattern designation and simply repeats the pattern specified by the column spread. That's a savings in storage and increased performance—a win-win situation.

So why don't we compress in the row and column directions at the same time and get even more? In effect, that's what a quadtree data structure does. It's an interesting second cousin to the traditional raster data structure that uses a cascading set of grid resolutions to compress redundancy. Consider the map boundaries shown in figure 19.1. If the map window is divided in half in both the X and Y directions, four panels (quadrants) are identified. That superimposes a coarse grid of just two columns and two rows.

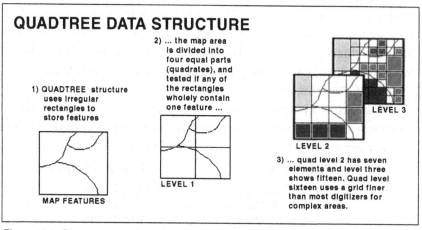

Figure 19.1. Quadtree data structure elements.

At that point the computer tests whether any quadrant wholly contains a single map characteristic. In the example there are none, so each of the panels is divided into its quads (Level 2). At that point, there are seven of the 16 quadrants that wholly contain a single characteristic. Their positions in the four-by-four grid are noted, and the remaining nine mixed panels are divided into their quads (Level 3). Fourteen of these are noted as completed, and the remaining 22 are divided for Level 4 of the quadtree. The process is repeated until an appropriate quad level resolution is reached. At level 16, a gridding resolution of 65,536 by 65,536 is available wherever it is needed—that equates to a fixed raster grid of 4,294,967,296 cells! But the quadtree isn't forced to use that resolution everywhere so it accurately stores most maps in less than a megabyte.

Quadtrees and run-length structures are good at compressing raster data. However, they must be decompressed and then recompressed for most map analysis operations. Many GISs don't impose a compression routine within the data structure, but simply leave it to commercial hard disk compression packages. Such packages respond to data redundancy and optimize for disk head movement.

The Triangulated Irregular Network (TIN) data structure is a vector offshoot originally designed for elevation data. It avoids the redundancy of elevations in a normal raster representation and is more efficient for some terrain analysis operations, such as slope and aspect. It uses a set of irregularly spaced elevation measurements with intensive sampling in areas of complex relief and/or important features, such as ridges and streams. A bit of computer wizardry is applied to determine the network of triangular facets that best fits these data. Each facet has three inter-connected elevations and can be visualized as a tilted triangular plane. The direction cosines of the plane identify its slope and aspect. The average of the three elevations generalizes the plane's height.

As shown in figure 19.2, the XY coordinate (location) and the Z coordinate (elevation) are stored in a points table similar to traditional vector structure. The triangular facets are defined in a features table by their three nodes and adjoining facets. The final link is to an attribute table which contains descriptive information on each facet. Awesome shaded relief maps can be generated by plotting the facets in 3-D and shading them as functions of their slopes and aspects.

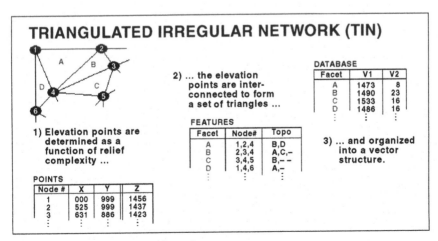

Figure 19.2. TIN data structure elements.

Using a TIN rather than raster structure to characterize a three-dimensional surface has some significant advantages. It usually requires fewer points, captures discontinuities such as streams and ridges, and determines slope and aspect of the facet itself. It's the data structure of choice for most civil engineering packages designed for terrain analysis. However, it's inappropriate for a generalized GIS that mixes a variety of maps. First, it's like raster, because it uses a mosaic of geographic chunks to represent a map feature. The chunks are inconsistent between maps, however, and something as simple as map overlay takes a severe hit in performance. Also, TIN two-dimensional renderings are complex and bewildering to most users compared to a normal vector plot.

Quadtree and TIN are useful offshoots of basic raster and vector data structures. They provide important benefits for certain data under certain conditions. If they match your needs, they're an invaluable addition to your GIS arsenal.

Rasterized Lines and Vectorized Cells

Offshoots of Traditional Raster and Vector Formats

20

Chances are your GIS is (or will be) ambidextrous. It has a vector side and a raster side, and might even have TIN or quadtree sides. The different data structures indicate various perspectives on data types and user applications. The vector approach characterizes discrete map objects and is influenced by applications in computer graphics. Raster characterizes continuous mapped data and emerges from remote sensing applications that involve multivariate statistics. Today, considerations in database/hardware structure influence future development as much as historical user/GIS representation theory. In a sense, the realities of an evolving computer environment challenge traditional ways.

A rasterized lines structure is an interesting offshoot from traditional data structures. It's sort of a hybrid because it uses a grid structure to characterize a map's line work. An optical scanner is used to "turn on" each cell in a fine sampling matrix that corresponds to the set of lines. The process is similar to the way your office fax machine "reads" a document. The fax at the other end simply deciphers the on-off conditions in the matrix and puts a dab of black toner at spots corresponding to "on." It skips over the "off" spots, leaving just white paper. That's it, a black-and-white rendering of map lines pushed over the phone lines.

If you use a magnifying glass, you can see the individual dots. At normal viewing distances, however, they merge to form smooth lines. Your brain easily makes sense of the pattern of lines and implied polygons embedded in the fine grid of the sampling matrix. Which streams are connected to which streams and which lakes are in which watersheds are obvious from the graphic rendering.

But that's not the case for a computer—the image is just a jumble of on-off dots. The first step in imposing order on data structure is to locate and mark as nodes all the special dots where lines meet (intersections), or just hang out by themselves (end points). Traditional vector structuring follows the dots between nodes, storing the coordinates for a point whenever there is a significant X or Y deflection. The result is a series of discrete points connected by implied straight lines, as shown in inset 2 of figure 20.1. The nodes and intervening points then are arranged as series of indexed files, as described in Chapter 19.

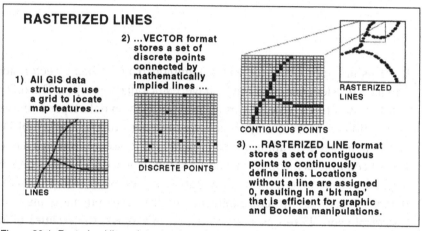

Figure 20.1. Rasterized lines data structure elements.

Rasterized lines retain all the dots along the lines. At first that seems stupid, because the points file is huge. Where an arc might require 10 discrete points, a rasterized line might require 100 or more. Some tricks in data compression and coordinate referencing can help, but it requires a lot more storage any way you look at it. So why would anybody use a rasterized line structure? Primarily because it's a format hardware loves. Faxes, scanners, screens, plotters and printers are based on dot patterns. As a result, much attention is paid to handling dotted data efficiently. Also, advances in optical disk storage and memory chips are redefining our concept of a "storage hog" so these files seem smaller each year.

More important, however, are advancements in central processing units (CPUs). Most computers still use a "kur-plunk-a" processing approach developed in the 1940s. It mimics our linear thinking and the way we do things—do A, then B, then C. Array and parallel processors, however, operate simultaneously on whole sets of data, *provided they're organized properly.*

Suppose you want to overlay a couple of traditional vector maps. In an instant, the computer reads the coordinates for two points on one map, then has to figure out if the implied line segment between them crosses any implied line segment on the other map. If it does, it mathematically calculates the point of intersection, splits the two lines into four, then updates all the indexed tables. Whew! That approach is a lot of work, and it's a purely linear process and a mismatch for array processing.

If the data are in rasterized line format, however, an array processor merely reads the corresponding chunks of cells on both maps and multiplies them together (a Boolean operation for you techy types). The product array identifies the composite of the two maps and highlights the new nodes where lines cross. Don't get me wrong: I'm not advocating you run out and buy rasterized line systems, but they have some interesting features for tomorrow's computers.

Another interesting offshoot, vectorized cells, cheats by storing each cell of an analysis grid as an individual polygon (figure 20.2). It just so happens that all of the polygons have the same square configuration and adjoin their neighbors. From a traditional vector perspective, each point defining a cell is a node; each cell side is an arc composed of just two nodes; and each cell is a polygonal feature composed of just four arcs. That approach uses the existing vector structure without impacting the existing code. It simply imposes consistency and uniformity in the polygons. All that's needed is an import module to generate the appropriate configuration of vectorized cells and read the raster values into a field in the corresponding attribute table. Raster-to-vector conversion can be completed by dissolving the pseudo-boundaries between adjoining polygons with the same value. Line features can be converted by connecting the centers of

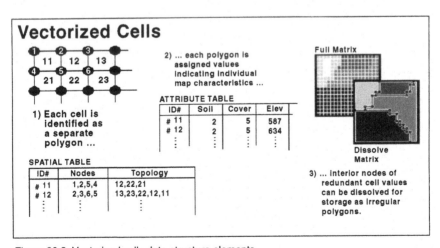

Figure 20.2. Vectorized cells data structure elements.

cells with similar value. Point features are represented by the coordinates of isolated cell centers.

Vectorized cells initially were used to "kludge" a link from raster to vector systems. However, the inherent consistency and uniformity of the structure (four points to a polygon), coupled with advances in data compression and database technology, might spark a resurgence in interest. Things as fluid as computers are becoming less bounded by storage requirements and the industry is shifting toward new processors and operating systems. Keep in mind that there are only two things certain about data structures: (1) tomorrow there will be new ones, and (2) what's good for one application isn't necessarily the best for another.

Recommended Reading

Books

Berry, J.K. "Introduction." in *Beyond Mapping: Concepts, Algorithms, and Issues in GIS*, Fort Collins, CO: GIS World Books, 1993.

Burrough, P.A. "Data Structures for Thematic Maps" and "Digital Elevation Models." Chapts. 2 and 3 in *Principles of Geographical Information Systems for Land Resources Assessment*, Oxford, UK: Clarendon Press, 1986.

Egenhofer, M.J., and J.R. Herring. "High-Level Spatial Data Structures for GIS." In *Geographical Information Systems: Principles and Applications*, ed. D.J. Maguire, M.F. Goodchild, and D.W. Rhind, Vol 1, 227-37. Essex, UK: Longman, 1991.

Franklin, W.R. "Computer Systems and Low-Level Data Structures for GIS," in *Geographical Information Systems: Principles and Applications*, ed. D.J. Maguire, M.F. Goodchild, and D.W. Rhind, Vol. 1, 215-25. Essex, UK: Longman, 1991.

Raper, J.F., and B. Kelk. "Three-Dimensional GIS," in *Geographical Information Systems: Principles and Applications*, ed. D.J. Maguire, M.F. Goodchild, and D.W. Rhind, Vol. 1, 299-317. Essex, UK: Longman, 1991.

Samnet, H. "Introduction," in *The Design and Analysis of Spatial Data Structures*, Redding, MA: Addison-Wesley, 1990.

Star, J., and J. Estes. "Data Structures." Chapt. 4 in *Geographic Information Systems: An Introduction*, Englewood Cliffs, NJ: Prentice Hall, 1990.

Tomlin, C.D. "Data." Chapt. 1 in *Geographic Information Systems and Cartographic Modeling*. Englewood Cliffs, NJ: Prentice Hall, 1990.

Journal Articles

Armstrong, M.P. and P.J. Densham. "Database Organization Alternatives for Spatial Decision Support Systems." *International Journal of Geographical Information Systems* 3(1): 27-33 (1990).

Gahegan, M.N., and S.A. Roberts. "An Intelligent, Object-Oriented Geographical Information System." *International Journal of Geographical Information Systems* 2: 101-10 (1988).

Mason, D.C., M.A. O'Conaill, and S.B Bell. "Handling Four-Dimensional Geo-Referenced Data in Environmental GIS." *International Journal of Geographical Information Systems* 8: 191-215 (1994).

Miline, P., S. Milton and J.L. Smith. "Geographical Object-Oriented Databases: A Case Study." *International Journal of Geographical Information Systems* 7(1): 39-55 (1993).

Worboys, M.F. "Object-Oriented Approaches to Geo-Referenced Information." *International Journal of Geographical Information Systems* 8(4): 385-99 (1994).7(1):39-55 (1993).

Worboys, M.F. "Object-Oriented Approaches to Geo-Referenced Information." *International Journal of Geographical Information Systems*, 8(4):385-99 (1994).

ORGANIZING THE MAP ANALYSIS TOOLBOX

FUNDAMENTAL COMPONENTS AND CONSIDERATIONS

They laughed at Joan of Arc,
but she went ahead and built it.
— Gracie Allen

What a GIS can do depends on the depth of the spatial information available to the computer, tempered by the depth of understanding of the analytical operations by those who use it. This section explains spatial topology and its extension to the classification of analytical GIS operations.

Spatial Topology

21

Differences among Graphics, Mapping, Spatial Database Management Systems, and GIS Approaches

To a human, a map is an image composed of colorful symbols. When you see a couple of red lines cross, your graphical intuition says, "a road intersection." When two blue lines combine into one, you think, "fork in a stream." As your eyes wander across a soil map, you easily grasp which soil unit is adjacent to which. Such truths are self-evident.

But that's not the case for a computer-compatible map. To the computer, a map is simply an organized set of numbers—no colored lines, no patterned globs. All of the relationships among map features must be captured in the number set, or the computer can't "see" the map. The term *spatial topology* describes the concept of this linkage, and can be thought of as information added to the pile of map coordinates.

Take a look at the map of the United States shown in figure 21.1. It's easy for you to detect the characteristic bumps for Florida, New England, and Texas. But the computer only sees thousands of "on-and-off" dots. If an individual dot is on, the computer assigns the appropriate color; it's totally unaware, however, of any patterns formed. This myopic rendering is characteristic of a *graphics* package. They're great for paint-

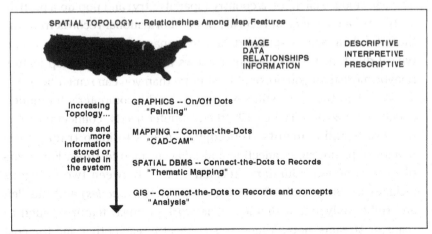

Figure 21.1. Spatial topology indicates the degree to which relationships among map features are known to the computer.

ing maps, but fail to offer the spatial topology needed for map analysis. A graphics package can't tell the difference between a map and the graphical rendering of a rose petal—both are just a pile of unrelated dots.

A *mapping* package is a bit more sophisticated, as it has "connect-the-dots" topology that outlines a distinct object. The data structure divides the set of all coordinates into piles, with a separate group for each distinct feature. One approach uses a "header" to identify the number of following coordinates that define the feature. If a point feature is indicated, only a pair of coordinates will follow. For a line feature, the header is followed by a string of coordinates connected sequentially. A polygonal feature marks a string of connected coordinates that closes on itself. That's the basic structure for an AutoCAD .DXF file—whether it's a blueprint for a sewage plant or a map of the world.

A spatial database management system extends this CAD-based structure to a "connect-the-dots-to-records" relationship. These packages link a CAD-like database, identifying the location of each map feature (spatial record), to another database containing information about each of the features (thematic records). The linkage is made through a common identification number (ID#) for each feature contained in the spatial and thematic datasets.

If you want to know which countries have a population greater than 200 million, the computer searches the appropriate field in the thematic database (thematic entry), then uses the ID#s to find the appropriate coordinates to draw each country that satisfies the query. Similarly, a user can "mouse-click" on a country (spatial entry) and pop up a particular record, a summary of records, or all informational records from the thematic database. A spatial database management system isn't your typical dumb map. The computer knows a lot about each map feature (maybe more than you do, or at least more than you can remember).

However, there are still several gaps in the computer's full understanding of the map. To be a GIS, the computer needs "connect-the-dots-to-records-and-concepts" topology. It needs to keep track of the relationships among connecting and adjacent map features. For example, the common boundary (termed an *arc*) between two polygons includes its "from and to" starting points (termed *nodes*) and the "left and right" polygons it divides. A network of linear features, such as roads or streams, notes which arcs connect to each other and the cost of traversing each arc in either direction. All this extra baggage of spatial topology does nothing to enhance the graphical rendering of a map; it merely gets in the way.

We go to all this trouble, however, because the computer can't find its way around on a nontopological map. A CAD-based road map might look good to you, but your computer sees a disorganized jumble of line segments. To determine an optimal path (or any path for that matter), the computer must have the connections you see stored in the dataset it manipulates. To determine the visual connectivity from one location to another, the computer needs to know the relative intervening elevations. To determine cover type diversity, it needs to quickly identify adjoining cover types around a location.

Each GIS package strikes a balance between stored and derived spatial topology. Vector systems tend to *store* a lot of their topology in the spatial tables linked to the thematic database. A simple "hit to disk" tells the computer the adjacent soil polygon or the next line segment along a road. Raster systems tend to *derive* their topology "on the fly" while processing the data. Finding an adjacent polygon or the next road cell involves a search of eight neighboring cells. In both vector and raster systems, intricate spatial relationships (e.g., point in polygon, intersecting lines, or effective buffers) are derived using the basic analytics in the GIS tool kit. Complex relationships involve spatial models containing several lines of code.

A GIS needs full spatial topology (connect the dots to records and concepts) to perform spatial analysis. As more information about the relationships among map features is bundled into the data structure or GIS tool set, the GIS can perform more work for you. If the system is kept in the dark, it can only draw a map—a simple picture of its database.

Like Nailing Jelly to a Tree 22

Classifying the Analytical
Capabilities of GIS

Classifying GIS analytical operations is a bit sticky. Tremendous inroads have been made toward a common understanding of data exchange formats, data structures, and even data content standards. However, agreement on a common, conceptual structure for GIS functionality remains elusive.

In part, that's due to the diverse disciplines claiming title to GIS and to their varied perspectives on what it should do. Coupled with these user differences is the vendor community's desire for product differentiation. The result is a quagmire in communicating GIS capabilities and freely exchanging application models.

Most GIS textbooks identify an essential set of GIS components as data input (encode), data management (store), manipulation/analysis (process), and product output (display). Discussions on the manipulation/analysis component tend to sort GIS operations into two broad categories: thematic and spatial. Thematic operations focus on *what*, or the attributes that describe map features. They involve processes such as data reclassification, aggregation, query, and conditional statements. For example, locating all of the management parcels (map features) containing Cohassett soil and Douglas fir trees (*what* attributes) involves a simple query to the management database, followed by a map display of the results.

Spatial operations focus on *where*, or location, and involve processing such as geometric translations, measurement, coincidence, and spatial statistics. These operations go beyond repackaging descriptive map data to creating entirely new spatial information and/or map features. For example, you could overlay a map of management parcels with a map of terrain steepness to derive an entirely new map identifying the average slope for each of the management parcels. As a result, you have new information (average slope) that didn't previously exist in the database. Or, the overlay could generate a new map with the management parcels partitioned into a subset of new map features based on the relative terrain steepness within the parcel.

At first, the distinction between thematic and spatial operations might seem trivial—merely semantics among the academics. However, the distinction is a major determinant of current GIS applications. Thematic operations reflect well-established database procedures that follow standard Structured Query Language (SQL) protocol. As a result, these applications have a large following of users within the greater computer community.

Spatial operations, however, present new concepts *and* foreign procedures. To a confused GIS-neophyte, there appear to be as many organizational schemes for spatial operations as there are GIS products and textbooks. However, there are a few common threads among the different taxonomies. First, they all differentiate spatial analysis from "housekeeping" (encoding and storage) and "visualization" (query and display). Second, they all agree that spatial analysis implies creating new mapped data—either new feature characteristics or new spatial partitioning.

The differences in organizational schemes tend to arise from the taxonomical structure itself—primarily a dichotomy between the developer and user camps. Developer-oriented schemes group the various spatial operations by how they work. This approach is well-suited for GIS developers, programmers, and specialists, because it relates to the algorithmic approaches ingrained in GIS processing. For example, Tomlin's comprehensive book on spatial analysis identifies three functional groups based on how the computer algorithm obtains mapped data for processing:[1]

1.   Local functions involve single or multiple values associated with *individual locations.*

2.   Focal and incremental functions involve values of immediate or extended *neighborhoods.*

3.   Zonal functions involve entire or partial zones, or *regions.*

User-oriented schemes, however, focus on input and output products. The approach is appropriate for general GIS users because it relates to familiar manual map processing procedures. My favorite identifies four functional groups:[2]

1.   Reclassification operations assign a *new value to each map feature* on a single map based on the feature's position, initial value, size, shape, or contiguity (clumps).

2.   Overlay operations assign new values *summarizing the coincidence of map features* from two or more maps based on a point-by-point, regionwide, or mapwide basis.

3.   Distance measurement assigns map values based on *simple or weighted connections among map features* including distance, proximity, movement, and connectivity (optimal paths, line-of-sight, and narrowness).

4.   Neighborhood operations assign map values that summarize *conditions within the vicinity of map locations* (roving window) based on surface configuration or statistical summary.

From a developer's perspective, calculating "average slope" for each management parcel is a zonal operation (summary of slope data within each parcel), whereas the "partitioning" of individual parcel/slope subdivisions is a local operation (intersecting vector lines or raster cells). From a user's perspective both are simply overlay operations that involve the coincidence of two maps. The distinctions arise because the developer relates to the differences in the two algorithms, while the user relates to manually superimposing the two maps on a light table.

A third perspective, application-orientation, also is used to organize spatial operations. For example, Environmental Systems Research Institute, Inc.'s GRID cell-based modeling toolkit contains more than 200 operations organized into 20 functional groups. The scheme draws from focal and zonal functions (reclassification and distance functions), and identifies application-specific groups to include geometric transformation, statistical, surface and shape analysis functions. Most of the groups, however, distinguish among mapematical operations to include arithmetic, Boolean, relational, bitwise, combinatorial, logical, accumulative, assignment, trigonometrical, exponential, and logarithmic.

Two things should be apparent: (1) we aren't clear about what GIS can do, and (2) we desperately need to be more clear. Before GIS can become a useful button on everyone's computer, there needs to be a level of consistency in processing structure that approaches what's being established in data structures. Without such consistency, we might be able to exchange data, but our spatial reasoning with the data will be fragmented and incomplete—a GIS Tower of Babel. Of course, data considerations aren't nailed down either. But that's another story.

1. Tomlin, C.D. *Geographic Information Systems and Cartographic Modeling*, Prentice-Hall Publishers, Englewood Cliffs, N.J., 1990.

2. Berry, J.K. "Cartographic Modeling: The Analytical Capabilities of GIS," in *Environmental Modeling with GIS*, Goodchild, Parks, and Steyaert, eds., pp. 58-74, Oxford University Press, Oxford, UK, 1993.

Resolving Map Detail 23
The Critical Factor of Map Resolution

What determines a map's accuracy? There are a lot of factors, but some important ones hinge on the concept of *resolution*. That's not a reference to the determination or tenacity of the cartographer, but a measure of the level of detail captured in a map. If a map captures more detail than another map, it has a higher (or finer) resolution.

In one sense, resolution can be related to map scale. We all know that more detail is seen in a map at 1:24,000 (large/local scale) than one of the same area at 1:2,000,000 (small/global scale). The effect is that we have only a few inches of space on a sheet of paper, and if each inch represents 2,000,000 feet (nearly 400 miles) on the ground, there isn't much room for details—hence, low resolution.

But scale only mathematically relates map measurements to actual ground distances. It doesn't fully account for the informational scale of a map. *Minimum mapping resolution* (MMR) notes the "level of spatial aggregation," which can be thought of as the smallest area that can be circled and called one thing. For example, the MMR for a 1:24,000 vegetation map is typically less than five acres. Sure you can discern a single tree, but would you circle it and call it a timber stand? What's it take—two trees, 10 trees ...?

The MMR for a 1:24,000 soils map is often six to 20 acres, with abundant disclaimers about possible "pockets" of other soils (globs of different soils smaller than the MMR). This informational scale is left to the discretion of the photo interpreter—largely a function of experience, the pen's width, air photo scale, and the discernability and homogeneity of the forest and soil units.

Another scale-related consideration is *spatial resolution*, identifying "the smallest addressable unit of space" used in delineating map features. In a vector system, the smallest addressable unit is the implied line segment connecting two points. If a point feature is denoted, the length of the line segment is zero, and the spatial resolution is at coordinate accuracy of the reference grid + digitizing error. As shown in figure

23.1, the spatial resolution of an arc is a function of the spacing of the digitized points—the closer the points, the higher the spatial resolution (especially on curved segments). A measure of the spatial resolution for a line involves the ratio of deflections in the X and Y directions to line segment length.

The spatial resolution for a raster system is simply the size of cell implied by the analysis grid—the smaller the cell, the higher the spatial resolution (see figure 23.1). Point features, such as a spring on a water map, are assumed to be contained in a single cell, with the minimal positional accuracy of one-half the diagonal of the cell.

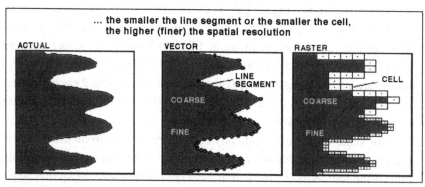

Figure 23.1. Spatial resolution identifies the smallest addressable unit of space. It's the *line segment* in a vector system, and it's the *cell* in a raster system.

Feature size and positioning aren't the only determinants of map detail. *Thematic resolution* identifies the smallest classification grouping of a map theme (see figure 23.2). In some applications, a simple forest/nonforest map might provide a sufficient description of vegetative cover. For years, this coarse classification has appeared as green on U.S. Geological Survey topographic sheets. Resource managers require a higher thematic resolution, however, and expand the classification scheme to include forest species, age and stocking level.

Another dimension of resolution, termed *temporal resolution*, identifies the frequency of map update. For example, a county planner might be content with a land-use map that's updated every couple of years. The farm agent for the county, however, needs the agricultural land-use theme broken into farm production classes (finer thematic resolution),

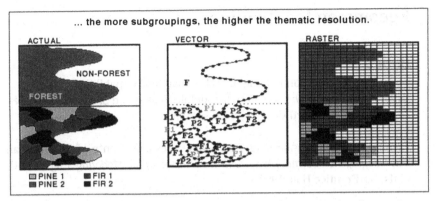

Figure 23.2. Thematic resolution identifies the smallest classification grouping of a map theme.

and these areas need to be updated a couple of times each year (finer temporal resolution).

The concept of informational scale is important in GIS database design. A corporate database requires consistency among its mapped data, or at least specification and translation procedures to track and adjust for inconsistencies. That's a far cry from the traditional plethora of personal paper maps.

For 8,000 years, geographic scale has been the de facto indicator of map detail. But times have changed, and measures of mapping, spatial resolution, thematic resolution and temporal resolution should be integral parts of the modern map's legend and processing procedures. Just keep in mind, the next time your GIS slams a few maps together, that simply translating to the same geographic scale and projection doesn't ensure consistent informational scales. And we all know what happens when you mix scales (ahhhhha!).

Recommended Reading

Books

Berry, J.K. "Cartographic Modeling: The Analytical Capabilities of GIS," in *Environmental Modeling with GIS*, Goodchild, Parks, and Steyaert (eds.), 58-74. Oxford, UK: Oxford University Press, 1993.

Tomlin, C.D. "Data Processing" and "Data Processing Control," Chapts. 2 and 3 in *Geographic Information Systems and Cartographic Modeling*. Englewood Cliffs, NJ: Prentice Hall, 1990.

Journal Articles

Berry, J.K. "Fundamental Operations in Computer-Assisted Map Analysis." *International Journal of Geographical Information Systems* 1: 119-36 (1987).

Persson, J., and E. Jungert. "Generation of Multi-Resolution Maps from Run-Length-Ecoded Data." *International Journal of Geographical Information Systems* 6(6): 497-510 (1992).

THE ANATOMY OF A GIS MODEL

SOME CASE STUDIES

> He that will not apply new
> remedies must expect new evils.
> — *Francis Bacon*

It is often stated that a GIS is only as good as its data. But in reality, it is only as good as the expression of its data—a refinement that includes GIS modeling—as well as the exactness of the database. This section compares several GIS models to illustrate different modeling approaches and the varying levels of results they generate.

From Recipes to Models

Basic Binary and Rating Model Expressions

24

So what's the difference between a recipe and a model? Both seem to mix a bunch of things together to create something else. Both result in a synergistic amalgamation that's more than the sum of the parts. Both start with basic ingredients and describe the processing steps required to produce the desired result—be it a chocolate cake or a landslide susceptibility map.

In a GIS, the ingredients are base maps and the processing steps are spatial handling operations. For example, a simple recipe for locating landslide susceptibility involves ingredients such as terrain steepness, soil type, and vegetation cover; areas that are steep, unstable, and bare are the most susceptible.

Before computers, identifying areas of high susceptibility required tedious manual map analysis procedures. A transparency was taped over a contour map of elevation, and areas where contour lines were spaced closely (steep) were outlined and filled with a dark color. Similar transparent overlays were interpreted for areas of unstable soils and sparse vegetation from soil and vegetation base maps. When the three transparencies were overlaid on a strong light source, the combination was deciphered easily—clear = not susceptible, and dark = susceptible. That basic recipe has been with us for a long time. Of course, the methods changed as modern drafting aids replaced the thin parchment, quill pens, and stained glass windows of the 1800s, but the conceptual approach remains the same.

In a typical vector GIS, a logical combination is achieved by first generating a topological overlay of the three maps (SLOPE, SOILS, COVERTYPE), then querying the resultant table (TSV_OVL) for susceptible areas. The Structured Query Language (SQL) query might look like the following:

Select columns:	%slope, stability, vegtype
from tables:	TSV_OVL
where condition:	%slope > 13 AND stability = "Unstable" AND vegtype = "Bare"
into table named:	L_HAZARD

The flowchart in figure 24.1 depicts a typical raster-based binary model (only two states of either Yes or No), which mimics the manual map analysis process and achieves the same result as the overlay/SQL query. A slope map is created by calculating the change in elevation throughout the project area (first derivative of the elevation surface). A

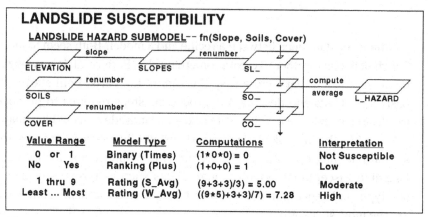

Figure 24.1. Binary, ranking, and rating models of landslide susceptibility. The location indicated by the piercing arrow contains a 34 percent slope, a fairly stable soil, and sparse forest cover.

binary solution codes as 1 all of the susceptible areas on each of the factor maps (>30 percent slope, unstable soils, bare vegetative cover), whereas the nonsusceptible areas are coded as 0. The product of the three binary maps (SL_*binary*, SO_*binary*, CO_*binary*) creates a final map of landslide potential—1 = susceptible, and 0 = not susceptible. Only locations susceptible on all three maps retain the "susceptible" classification (1*1*1= 1). In the other instances, multiplying 0 times any number forces the product to 0 (not susceptible). The mapematical model corresponding to the flowchart in figure 24.1 might be expressed (in TMAP modeling language) as

SLOPE ELEVATION FOR SLOPES

RENUMBER SLOPES FOR SL_binary ASSIGNING 0 TO 1 THRU 12

    ASSIGNING 1 TO 13 THRU 1000   (> 13%, steep)

RENUMBER SOILS FOR SO_binary ASSIGNING 0 TO 0 THRU 2

    ASSIGNING 1 TO 3 THRU 4   (soils 3&4, unstable)

RENUMBER COVERTYPE FOR CO_binary ASSIGNING 0 TO 1

    ASSIGNING 0 TO 3 ASSIGNING 1 TO 2   (cover 2, bare)

COMPUTE SL_binary TIMES SO_binary

    TIMES CO_binary FOR L_HAZARD   (1*1*1=1, hazard)

In the multiplicative case, the arithmetic combination of the maps yields the original two states—dark or 1 = susceptible, and clear or 0 = not susceptible. It's analogous to the "AND" condition of the logical combination in the SQL query. However, other combinations can be derived. For example, the visual analysis could be extended by interpreting the various shades of gray on the stack of transparent overlays: clear = not susceptible, light gray = low susceptibility, medium gray = moderate susceptibility, and dark gray = high susceptibility. In an analogous mapematical approach, the computed sum of the three binary maps yields a similar ranking: 0 = not susceptible, 1 = low susceptibility, 2 = moderate susceptibility, and 3 = high susceptibility (1+1+1 = 3). That approach is called a ranking model, because it develops an ordinal scale of increasing landslide potential—a value of two is more susceptible than a value of 1, but not necessarily twice as susceptible.

A rating model is different, because it uses a consistent scale with more than two states to characterize the relative landslide potential for various conditions on each factor map. For example, a value of 1 is assigned to the least susceptible steepness condition (e.g., from 0 percent to 5 percent slope), while a value of 9 is assigned to the most susceptible condition (e.g., > 30 percent slope). The intermediate conditions are assigned appropriate values between the landslide susceptibility extremes of 1 and 9. That calibration results in three maps with relative susceptibility ratings (SL_rate, SO_rate, CO_rate) based on the 1-9 scale.

Computing the simple average of the three rate maps determines an overall landslide potential based on the relative ratings for each factor at each map location. For example, a particular grid cell might be rated 9, because it's steep, 3 because its soil is fairly stable, and 1 because it's forested. The average landslide susceptibility rating under these conditions is $[(9+3+3)/3] = 5$, indicating a moderate landslide potential.

A weighted average of the three maps expresses the relative importance of each factor to determine overall susceptibility. For example, steepness might be identified as five times more important than either soils or vegetative cover in estimating landslide potential. For the example grid cell described previously, the weighted average computes to $[([9*5]+3+3)/7] = 7.28$, which is closer to a high overall rating. The weighted average is influenced preferentially by the SL_rate map's high rating, yielding a much higher overall rating than the simple average.

All that may be a bit confusing. The four different "recipes" for land-slide potential produced strikingly different results for the example grid cell in figure 24.1—from not susceptible to high susceptibility. It's like baking banana bread. Some folks follow the traditional recipe, some add chopped walnuts or a few cranberries. By the time diced dates and can-died cherries are tossed in, you can't tell the difference between your banana bread and last years' fruitcake.

So back to the main point—what's the difference between a recipe and a model? Merely semantics? Simply marketing jargon? The real dif-ference between a recipe and a model isn't in the ingredients, nor the processing steps themselves. It's in the conceptual fabric of the process. But more on that later.

It's All Downhill from Here

Extending Basic Models Through Logic Modifications

25

Chapter 24 described various renderings of a landslide susceptibility model. It related the results obtained for an example location using manual, logical combination, binary, ranking, and rating models. The results ranged from not susceptible to high susceptibility. Two factors in model expression were at play: the type of model and its calibration. However, the model structure, which identified the factors considered and how they interact, remained constant. In the example, the logic was constrained to jointly considering terrain steepness, soil type, and vegetation cover. One could argue other factors might contribute to landslide potential. What about depth to bedrock? Or previous surface disturbance? Or slope length? Or precipitation frequency and intensity? Or gopher population density? Or about anything else you might dream up?

That's it. You've got the secret to seat-of-the-pants GISing. First you address the critical factors, then extend your attention to other contributing factors. In the abstract it means adding boxes and arrows to the flowchart to reflect the added logic. In practice it means expanding the GIS macro code, and most importantly wrestling with the model's calibration. For example, it's easy to add a fourth row to the landslide flowchart, identifying the additional criterion of depth to bed rock, and tie it to the other three factors. It's even fairly easy to add the new lines of code to the GIS macro:

RENUMBER DEPTH_BR FOR BR_**binary**

ASSIGNING **0** TO **0** THRU **4**

ASSIGNING **1** TO **5** THRU **15**   (>4, massive)

COMPUTE SL_binary TIMES SO_binary

TIMES CO_binary **TIMES BR-BINARY** FOR L_HAZARD

Things get a lot tougher when you have to split hairs about precisely what soil depths increase landslide susceptibility (>4 a good guess?).

The previous discussions focused on the hazard of landslides, but not their risk. Do we really care about landslides unless there is something valuable in the way? Risk implies the threat a hazard imposes on something valuable. Common sense suggests that a landslide hazard distant from important features represents a much smaller threat than a similar hazard adjacent to a major road or school. Figure 25.1 shows a risk extension to the rating model that considers proximity to important features as a risk indicator. In the flowchart, a map of proximity to roads (R_PROX) is generated that identifies the distance from every location to the nearest road. Increasing map values indicate locations farther from a road. A binary map of buffers around roads (R_BUFFERS) is created by renumbering the distance values near roads to 1 and far from roads to 0. By multiplying this "masking" map by the landslide susceptibility map (hazard), the landslide threat is isolated for just the areas around roads (risk).

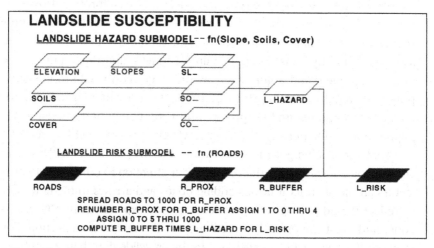

Figure 25.1. Risk extension to landslide susceptibility model.

A further extension to the model involves variable width buffers as a function of slope (figure 25.2). The logic in that refinement is that in steep areas the buffer width increases as a landslide poses a greater threat. The threat diminishes in gently sloped areas, so the buffer width contracts. The weighted buffer extension calibrates the slope map into an impedance map (FRICTION), which guides the proximity measurement. As the computer calculates distance in steep areas (low impedance), it assigns larger effective distance values for a given geographic step than it does in gently sloped areas (high impedance). That results in an effective proximity map (R_WPROX), with increasing values indicating locations that are effectively farther away from the road. The buffer map from

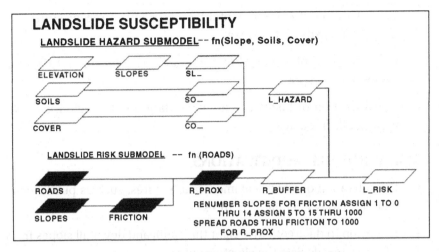

Figure 25.2. Weighted buffer extension to landslide susceptibility model.

these data is radically different from the simple buffer in the previous model extension. Instead of a constant geographic reach around the roads, the effective buffer varies in width, as a function of slope, throughout the map area. As before, the buffer can be used as a binary mask to isolate the hazards within the variable reach of the roads.

That iterative refinement characterizes a typical approach to GIS modeling—from simple to increasingly complex. Most applications first mimic manual map analysis procedures and are then extended to include more advanced spatial analysis tools. For example, a more rigorous mapematical approach to the previous extension might use a mathematical function to combine the effective proximity (R_WPROX) with the relative hazard rating (L_HAZARD) to calculate a risk index for each location. For your enjoyment, some additional extensions are suggested below. Can you modify the flowchart to reflect the changes in model logic? If you have tMAP, can you develop the additional code? If you're a malleable undergraduate, you have to if you want to pass the course. But if you're a professional, you need not concern yourself with such details. Just ask the 18-year-old GIS hacker down the hall to do your spatial reasoning.

HAZARD SUBMODEL MODIFICATIONS

- Consideration of other physical factors, such as bedrock type, depth to bedrock, faulting, etc.

- Consideration of disturbance factors, such as construction cuts and fills

- Consideration of environmental factors, such as recent storm frequency, intensity and duration

- Consideration of seasonal factors, such as freezing and thawing cycles in early spring

- Consideration of historical landslide data for the region, such as earthquake frequency

RISK SUBMODEL MODIFICATIONS

- Consideration of additional important features, such as public, commercial, and residential structures

- Extension to differentially weight the uphill and downhill slopes from a feature to calculate the effective buffer

- Extension to preferentially weight roads based on traffic volume, emergency routes, etc.

- Extension to include an economic valuation of threatened features and potential resource loss

Due Processing 26
Evaluating Mapematical Relationships

As noted in Chapters 24 and 25, GIS applications come in a variety of forms. The differences aren't as much in the ingredients (maps) or the processing steps (command macros) as in the conceptual fabric of the process. As noted in the evolution of the landslide susceptibility model, differences can arise through model logic and/or model expression. A simple binary susceptibility model (only two states of either Yes or No) is radically different from a rating model using a weighted average of relative susceptibility indices. In mathematical terms, the rating model is more robust, because it provides a continuum of system responses. Also, it provides a foothold to extend the model even further.

There are two basic types of GIS models: cartographic and spatial. In short, a cartographic model focuses on automating manual map analysis techniques and reasoning, and a spatial model focuses on expressing mathematical relationships. In the landslide example, the logical combination and the binary map algebra solutions are obviously cartographic models. Both could be manually solved using file cabinets and transparent overlays—tedious, but feasible for the infinitely patient. The weighted average rating model, however, smacks of down and dirty mapematics and looks like a candidate spatial model. But is it?

As with most dichotomous classifications there is a gray area of overlap between cartographic and spatial model extremes. If the weights used in rating model averaging are merely guess-timates, then the application lacks all of the rights, privileges, and responsibilities of an exalted spatial model. The model may be mathematically expressed, but the logic isn't mathematically derived, nor empirically verified. In short, "Where's the science?"

One way to infuse a sense of science is to perform some data mining. That involves locating a lot of areas with previous landslides, then pushing a predictive statistical technique through a stack of potential driving variable maps. For example, you might run a regression on landslide occurrence (dependent mapped variable) with %slope, %clay, %silt, and

%cover (independent mapped variables). If you get a good fit, then substitute the regression equation for the weighted average in the rating model. That approach is at the threshold of science, but it presumes your database contains just the right set of maps over a large area. An alternative is to launch a series of "controlled" experiments under various conditions (%slope, soil composition, cover density, etc.) and derive a mathematical model through experiment. That's real science, but it consumes a lot of time, money, and energy.

A potential shortcut involves reviewing the scientific literature for an existing mathematical model and using it. That approach is used in figure 26.1, a mapematical evaluation of the Revised Universal Soil Loss Equation (RUSLE)—kind of like landslides from a bug's perspective. The expected soil loss per acre from an area, such as a farmer's field, is determined from the product of six factors: the rainfall, the erodibility of the soil, the length and steepness (gradient) of the ground slope, the crop grown in the soil, and the land practices used. The RUSLE equation and its variable definitions are shown in figure 26.1. The many possible numerical values for each factor require extensive knowledge and preparation. However, a soil conservationist normally works in a small area, such as a single county, and often needs only one or two rainfall factors (R), values for only a few soils (K), and only a few cropping/practices systems (C and P). The remaining terrain data (L and S) are tabulated for individual fields.

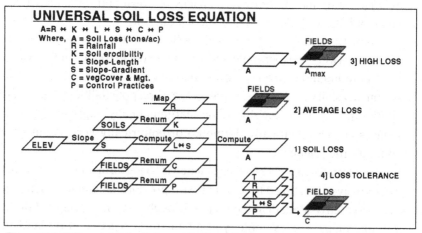

Figure 26.1. GIS model of the Universal Soil Loss Equation.

The RUSLE model can be evaluated two ways: aggregated or disaggregated. An aggregated model uses a spatial database management system (DBMS) to store the six factors for each field, then solves the equation through a database query. A map of predicted soil loss by individual field can be displayed, and the total loss for an entire watershed can be calculated by summing each of the constituent field losses (loss per acre times number of acres). That RUSLE implementation provides several advancements, such as geo-query access, automated acreage calculations, and graphic display, over the current procedures.

However, it also raises serious questions. Many fields don't fit the assumptions of an aggregated model. Field boundaries reflect ownership rather than uniformly distributed RUSLE variables. Just ask any farmer about field variability (particularly if their field's predicted soil loss puts them out of compliance). A field might have two or more soils, and it might be steep at one end and flat on the other. Such spatial variation is known to the GIS (e.g., soil and slope maps), but not used by the aggregated model. A disaggregated model breaks an analysis unit (farmer's field in this case) into spatially representative subunits. The equation is evaluated for each of the subunits, then combined for the parent field.

In a vector system, the subunits are derived by overlaying maps of the six RUSLE factors, independent of ownership boundaries. In a raster system, each cell in the analysis grid serves as a subunit. The equation is evaluated for each "composite polyglet" or "grid cell," then weight-averaged by area for the entire field. If a field contains three different factor conditions, the predicted soil loss proportionally reflects each subunit's contribution. The aggregated approach requires the soil conservationist to fudge the parameters for each of the conditions into generally representative values, then run the equation for the whole field. Also, the aggregated approach loses spatial guidance for the actual water drainage—a field might drain into two or more streams in different proportions.

Figure 26.1 shows several extensions to the disaggregated model. Inset 1 depicts the spatial computations for soil loss. Inset 2 uses field boundaries to calculate the average soil loss for each field based on its subunits. Inset 3 provides additional information not available with the aggregated approach. Areas of high soil loss (AMAX) are isolated from the overall soil map (A), then combined with the FIELDS map to locate areas out of compliance. That directs the farmer's attention to portions of the field which might require different management action. Inset 4 enables the farmer to reverse calculate the RUSLE equation. In this case, a soil loss tolerance (T) is established for an area, such as a watershed, then the combinations of soil loss factors meeting the standard are derived. Because the climatic and physiographic factors of R, K, L, and S are beyond a farmer's control, attention is focused on vegetation cover (C) and control practices (P). In short, the approach generates a map of the set of crop and farming practices that keep the field within soil loss compliance—good information for decision making.

OK, what's wrong with the disaggregated approach? Two things: our databases and our science. For example, our digital maps of elevation may be too coarse to capture the subtle tilts and turns that water follows. And the science behind the RUSLE equation may be too coarse (modeling scale) to be applied to quarter-acre polyglets or cells. These limitations, however, tell us what we need to do—improve our data and redirect our science. From that perspective, GIS is more of a revolution in spatial reasoning than an evolution of current practice into a graphical form.

Author's Note: Let me apologize for this brief treatise on an extremely technical subject. How water cascades over a surface, or penetrates and loosens the ground, is directed by microscopic processes. The application of GIS (or any other expansive model) by its nature muddles the truth. The case studies presented are intended to illustrate various GIS modeling approaches and stimulate discussion about alternatives.

Recommended Reading

Books

Berry, J.K. "Measuring Effective Distance and Connectivity" and "Cartographic and Spatial Modeling." Topics 2 and 10 in *Beyond Mapping: Concepts, Algorithms, and Issues in GIS*, Fort Collins, CO: GIS World Books, 1993.

Berry, J.K., and W. Ripple. "Emergence and Role of GIS in Natural Resource Information Systems," in *The GIS Applications Book: Examples in Natural Resources*, 3-20. Falls Church, VA: American Society of Photogrammetry and Remote Sensing, 1994.

Goodchild, M.F. (ed.). "From Modeling to Policy." Topic 4 in *Environmental Modeling with GIS*, ed. M.F. Goodchild, B.O. Parks, and L.T. Steyaert, Oxford, UK: Oxford University Press, 1993.

Tomlin, C.D. "Descriptive Modeling" and "Prescriptive Modeling." Chapts. 7 and 8 in *Geographic Information Systems and Cartographic Modeling*. Englewood Cliffs, NJ: Prentice Hall, 1990.

Journal Articles

Pereira, J.M.C., and L. Duckstein. "A Multiple Criteria Decision-Making Approach to GIS-Based Land Suitability Evaluation." *International Journal of Geographical Information Systems* 7(5): 407-24 (1993).

PUTTING GIS IN THE HANDS OF PEOPLE

CONSIDERATIONS AND COMPONENTS OF A FIELD UNIT

Fast, good, cheap... pick two.
— Anonymous

GIS is but one component of the spatial sciences, focusing on the relationships within and among mapped data. The allied field of Global Positioning Systems (GPS) focuses on precise positioning in space while remote sensing technology focuses on monitoring and classifying the landscape. This section discusses the underlying principles of these related fields and describes their integration into a GIS/GPS/remote sensing field unit.

Putting Things in Their Proper Places 27

The Basics of Global Positioning Systems

GIS is awesome. It allows you to view maps in the blink of an eye, visualize spatial patterns in datasets, and even model complex relationships among mapped variables. But its abstract renderings (i.e., maps) require a real-world expression to make GIS a practical tool. For years, "field folk" have been swatting mosquitoes and lusting for a simple way to figure out where they are and where they're going. Celestial navigation methods used by early mariners as they gazed at the heavens eventually gave way to surveying and mapping sciences. But such solutions still seem beyond the grasp of average bushwhackers. What's needed is a simple field unit that puts the awesome power of GIS in their hands.

That's where the Global Positioning System (GPS) comes in. Based on a constellation of 21 satellites, each of which circles the globe every 12 hours, GPS links GIS maps and their datasets to real-world positions and movements. In effect, a set of man-made stars supports the electronic equivalent of celestial navigation. How does it work? And will it work for you?

Figure 27.1 shows important GPS considerations. The system employs the same principle of triangulation commonly used in scout camp and high school geometry. Circles of a calculated distance are drawn about a set of satellites whose precise positions are known through orbit mathematics described in satellite almanacs. The intersection of the circles determines your position on Earth. In trigonometric theory, only three channels (satellites) need to be monitored, but in practice four or more are required to cancel receiver clock errors. The radii of the circles are determined by noting the time lag for a unique radio signal from a satellite to reach you, then multiplying the elapsed time by the speed of light at which the radio waves travel. The world of electronic wizardry (involving technical stuff like pseudo-random code, carrier phase, ephemeris adjustments, and time hacks) allows timing to one billionth of a second (.0000000001), producing extremely accurate distance measurements in three-dimensional space. Generally, averaged

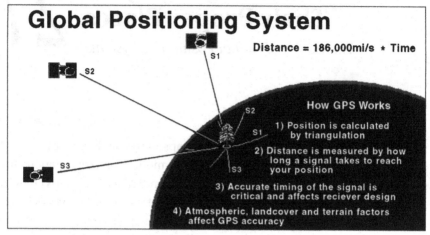

Figure 27.1. The basic elements of the Global Positioning System.

stationary measurements (static mode) tend to be more accurate than those made while moving (kinematic mode).

As with everything in the real world, actual GPS performance depends on several muddling factors, the most influential of which is GPS's history as a U.S. Department of Defense program. The government financed the $10 billion system for military purposes and feels uncomfortable if just anyone, such as terrorists or enemy troops, can simply tap into the system. The government purposely degrades the signal by using an operational mode, Selective Availability (S/A), which provides only 100-meter accuracy. With S/A turned off, 30-meter accuracy is common.

The degraded signal can be improved by a differential correction. A differential GPS uses a local reference receiver whose exact position is known. When the reference receiver gets a satellite signal, it calculates its implied position, then quickly "reverse calculates" the error correction needed to place it where it should be. The correction is broadcast to field units or stored for post-processing back at the office.

In general, there are two main hurdles in processing GPS signals: jitters and jumps. As with any instrument, inherent error for a set of readings at a fixed location yields a jittery cluster of possible positions, termed the sphere of uncertainty. Also, satellites come and go with time; as one is dropped and another picked up, the positions often take a temporary jump. Processing software uses running and dampened averages of several readings to cope with jitters and jumps. Keep in mind that the silicon in all GPS receivers is about the same—creative software separates one receiver from another.

A well-tuned differential GPS system in static mode can place you within a meter horizontally and five meters vertically, while a simple autonomous system for $200 or so can place you somewhere within a ball park—that is if atmospheric, ground cover and terrain factors permit—signals deteriorate under dense forest canopy or at the bottom of steep canyons. Also, the satellites aren't always available in a nicely dispersed pattern. That means you need to plan to be in the field at the times the satellites' celestial charts dictate. (Try explaining that one to your field crew.) Finally, it's important to keep in mind that GPS isn't intended to fully replace conventional surveys. Rather, it augments cadastral records with real-time and real-world positioning.

GPS's ability to locate positions on Earth's surface rapidly and accurately is a powerful addition to GIS. For example, the boundary of a wildfire can be digitized quickly with a GPS simply by walking (or flying in a helicopter) over the burn's perimeter—putting the fire in the GIS while it's still hot. From a forester's perspective, the GPSed map can be overlaid on existing inventory information to quantify timber lost and plan for salvage logging and forest regeneration. From a wildlife biologist's perspective, the burned area can be translated into habitat loss estimates, affected animal populations and ecosystem recovery plans. That means the forester, biologist, and others can be locked in honest debate regarding accurate and fully integrated data within hours of a geographic event.

In addition, a GPS receiver can be attached to a vehicle to generate an accurate map of important features and roads en route to various locations. According to rangers working in the U.S. Forest Service's Rocky Mountain region, GPS has reduced the time they spend in the field 50 to 80 percent, with minimal crew instruction. For example, a two-man team using GPS completed a section subdivision in less than a day—a task that normally takes a week with conventional survey techniques.

GPS's contribution to generating and updating GIS maps is obvious. Yet GPS is more than a data collection device—it's a practical tool to navigate GIS results. As GIS matures, more of its applications will involve GIS modeling, such as variable-width buffers around streams that allow for terrain steepness, ground cover, and soil erodibility. Although such relationships follow common sense, their spatial expression is complex. The contractions and expansions of a variable-width buffer on a paper map are practically gibberish to a field crew. If the coordinates of the buffer are loaded into a GPS, however, the result delineates the spatial reasoning and its complicated expression in the actual landscape.

A Lofty Marriage 28

The Basics of Remote Sensing Technology

As noted in Chapter 27, GIS/GPS technology positions spatial data and reasoning on the landscape. But effectively identifying, measuring, and monitoring the landscape over extensive areas is an ongoing challenge. Remote sensing, closely related to GIS, greatly enhances the technical mapping toolkit. Remote sensing is GIS's older brother, using relative variations in electromagnetic radiation (EMR) to identify landscape characteristics and conditions. In fact, so do your eyes. Sunlight, the form of EMR we see, starts off with fairly equal parts of blue, green, and red light. When sunlight is reflected from a leaf, the red and blue light is absorbed in photosynthesis and your eyes detect mostly the unused green light. Your brain interprets the subtle differences in the amount of blue, green, and red light to recognize the thousands of colors we relate to our surroundings.

A remote sensing satellite operates similarly. Its instruments focus for an instant at a spot on the ground measuring less than a quarter acre (see figure 28.1). Like your eyes, it records the relative amounts of the different types of light it "sees"—a lot of green for a dense, healthy forest; less green and more blue and red for bare ground. In addition to normal light (the

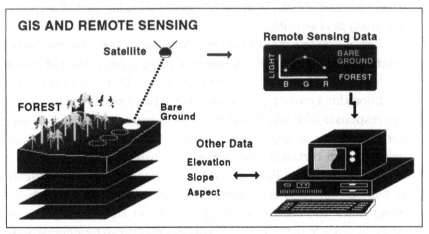

Figure 28.1. Remote sensing's basic elements.

visible spectrum), it can record other types of EMR our eyes can't see, such as near infrared, thermal, and radar energy. As the sensor sweeps from side to side and the satellite moves forward, it records the light from millions of spots, termed pixels for "picture elements." When the pixels are displayed on a computer, they form an image similar to an aerial photograph. Yet keep in mind that behind the image are millions of numbers that record the various types of light at each spot.

That organized mountain of numbers is used to identify land-cover characteristics and their condition. First, the computer is "trained " by locating representative areas of the cover types to be classified—sort of rubbing the computer's nose in what it should know. Then that information is used to classify other areas with similar EMR responses. As shown in the center of figure 28.2, the computer examines the amount of light for each type from the hundreds of training pixels in the examples. It notes that forests tend to have high green and low red responses, while bare ground has low green and slightly more red. The big dot in the center of each data cluster is the average amount of green and red light— the typical response for that cover type. Now the computer can consider the green/red responses for the millions of other locations and classify them through "guilt by association" with the training set statistics. In effect, through complex computer mathematics it plots an unknown location's green/red numbers (the "x" in the right graph of figure 28.2), notes the distance to the typical responses, then classifies the location as the closest cover type. Then it moves to the next spot, and the next ... until the entire area has been classified. You could do that, but your patience would ebb at about the second location for a set of several million in a typical satellite image.

Just as you use more than color to identify a tree, so can the computer. That's where GIS lends remote sensing a hand. The GIS uses the example locations to check its database to see if there are other typical conditions for a cover type. For example, if two forest types have similar EMR responses, the knowledge that the unknown location is "a steep, northerly slope at high elevation" might be enough to tip the scales toward a correct classification between the two.

In return for its help, the GIS gets a copy of the classification results—a completed cover map. By comparing the maps from two different times, the computer can quickly detect and quantify any land-cover changes. In the GIS, data on wildlife activity can be summarized for

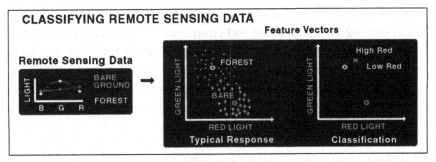

Figure 28.2. Classifying remote sensing data.

each cover type to see which is preferred. Once a preference is established, the loss (or gain) in an animal's preferred habitat can be inferred, measured, and plotted at megahertz speed. Keep in mind that remote sensing and GIS provide educated guesses at actual characteristics, conditions, and relationships. They aren't perfect, but they provide powerful and compelling insights.

In fact, remote sensing provides an element that's unavailable with traditional mapping: uncertainty assessment. At the moment of classification, the computer knows which typical response is closest and how close it is. If it's very close, then you're fairly confident of the classification. As it gets farther away, you're less confident. Relating the closest distance to those of other possible cover types yields even more information—sort of a probability sandwich of a location's cover type. The next closest typical response identifies a second guess at classification, and how much farther away it is indicates the degree of confusion between the two possibilities.

If an unknown location sits halfway between the typical responses of two cover types, it's a toss-up. As an unknown's response moves closer to a typical response, classification certainty increases—maybe, maybe not. That's where things can get a bit confusing. Note the data patterns (dots for forest and crosses for bare ground) in the typical response plot in figure 28.2. The forest responses in its training set is fairly compact and circular, while the bare responses are more dispersed and elongated in the direction of the typical forest response.

The shape of the data cluster, as well as its positioning, provides even more information about classification accuracy. The more dispersed the pattern is, the less typical the typical response is. If the data have a trend (elongated), it means the data are more apt to be confused with other cover types in that direction. All this statistical stuff is contained in the training set's joint mean and covariance matrix—take my word for it, or go back for an advanced degree in multivariate statistics. The upshot is that remote sensing classification tells you what it thinks is at a location and honestly reports how well it's guessing. This shadow map of certainty is the cornerstone of thematic error propagation; without it GIS models just flap in the wind.

See also Topic 4, Chapters 10 and 11 (pages 53-59).

Heads-Up and Feet-Down Digitizing 29

Components of a GIS/GPS/RS Field Unit

Chapters 27 and 28 described the GIS/GPS/remote sensing mapping triad. GIS expresses relationships among maps; GPS links map coordinates to real-world locations; and remote sensing directly records and classifies current views of the landscape. For years, GIS and remote sensing have been the realm of specialists in segregated offices "down the hall and to the right." In part, the division between mapped data providers and users was technological. GIS and remote sensing are inherently complicated fields, with more than a smattering of statistics, mathematics, and computer science. Also, their data loading and processing demands required expensive, specialized equipment.

More recently, low-end computers have grown up with storage, processing, and display capabilities approaching those of expensive workstations sold just a couple of years ago. Concurrently, the user community is becoming more computer literate, at least in PC-based applications. Moreover, more users recognize the importance of spatial attributes in datasets. With all of these trends in place, why isn't integrated GIS/GPS/remote sensing in the hands of more people?

Part of the answer lies in cultural lags for providers and users. The providers are close to the complexities of spatial data and their analysis. As a result, the providers focus on a capabilities "toolbox" that can do anything. The users, however, know exactly what they want the toolbox to do—usually automating current tasks. Anything more is simply confusing and esoteric theory. Both groups reflect their professional cultures and somewhat divergent views of the environments and the applications of spatial technology.

Another part of the answer lies in the delivery of spatial technology. By their nature, maps are abstract renderings of real-world objects. In its least abstract form, spatial processing mimics cartographic concepts that aren't well understood by most potential users. As a result, such users have an uneasy feeling about maps— particularly if they're on a computer. So what's needed to melt these spatial cold pricklies into warm fuzzies?

A fully integrated GIS/GPS/remote sensing field unit would help. For example, figure 29.1 shows an aerial photo (remote sensing) as a backdrop registered to a road map (GIS). The large star near the center of the figure identifies the GPS unit's current position. Now a skeptical user sees the road behind him and the clump of trees to the left. The integrated presentation takes the abstraction out of mapping and inserts human experience.

Important features can be encoded by tracing them on the screen with the aerial photo as a guide (termed heads-up digitizing, because your head is tilted up toward the screen). Or, as in this case, feet-down digitizing can be done by walking the perimeter of a field of interest. The proverbial "farmers from Missouri" can actually experience the link between a map and the real world.

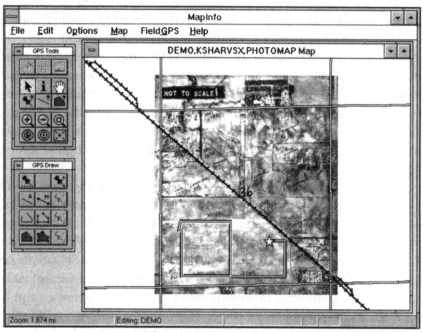

Figure 29.1. A GIS/GPS/remote sensing integrated display. Plot generated using MapInfo™ (MapInfo Corp.) and FarmGPS™ (Farmer's Software, Inc.).

In fact, a field GIS/GPS/remote sensing unit forms the foundation of precision farming, an emerging technology that places the unit in a tractor's cab and position stamps its location as it moves through a field. In a harvester, the unit can monitor crop yield and moisture content and relate measurement variances to soil maps and other terrain factors. On a spray rig, the unit can vary the application of chemicals as the tractor moves across a field. In crop research, the dataset represents a complete census of field inputs and outputs—a giant step beyond a few similar research plots at the university 70 miles away. Imagine the impact on any of the spatially oriented sciences and their management expressions.

So what comprises a GIS/GPS/remote sensing field unit? Chances are you'll need more than simple coasters attached to your GIS workstation and a long electrical cord. When "blue-skying" the possibilities with clients, I begin with four initial considerations: applications, operating environments, data loadings, and processing requirements. A clear, concise statement of what the device must do sets the stage for how it will be used and what capabilities are needed. For example, precision farming's application involves a mounted unit focused on precise kinematic GPS with extensive data loadings and on-board processing demands. A hand-held unit for timber inventory, however, focuses on a small set of field measurements, requiring minimal data storage and post-processing. But it all has to fit into a small, rugged shell.

Such considerations establish the scope of the application and its baseline requirements. The next step translates the requirements into the following design components:

- integrated software—GIS, GPS, remote sensing, statistics, office ...

- processor—'486, Pentium ...

- operating system/environment—UNIX, DOS, WINDOWS, NEXT ...

- memory—RAM, EPROM, flash cards, disk, tape ...

- ports/slots—RS232, LPT, PCMCIA, SCSI...

- screen—active/passive, reflective/transflective/backlighted ...

- input device—keyboard, pen, touch, voice ...

- peripherals—printer, plotter, sound, video, digital camera ...

- power—external/internal, protection, battery life, recharge rate ...

- durability—dust, splash, water, temperature, shock, electromagnetic fields ...

- physical—mounted/portable, size, weight, construction ...

Whew! That's a lot of techy stuff better left to the engineers (and there's a bunch of these new-wave GISers hard at work). The trend toward a GIS/GPS/remote sensing field unit promises to revolutionize current spatial technologies. No longer can the spatial triad operate independently. No longer can a one-size solution fit all applications. GIS's comprehensive toolbox needs to be open to other systems, reducible to the subset of directly needed functions and designed for small boxes—in short, tailored to individual end-user applications. GIS can't stop with the colorful plot of a map generated in the GIS office down the hall and to the right; it needs to be extended into the field and placed in the hands of people to support the spatial decisions they make and implement.

Recommended Reading

Books

Aronoff, S. "Remote Sensing." Chapt. 3 in *Geographic Information Systems: A Management Perspective*, Ottawa, Canada: WDL Publications, 1989.

Leick, A. "Introduction," "Elements of Satellite Surveying," "The Global Positioning System," and "Adjustment Computations," Chapts. 1, 2, 3, and 4 in *GPS Satellite Surveying*, New York, NY: John Wiley & Sons, 1990.

Star, J., and J. Estes. "Remote Sensing and GIS." Chapt. 10 in *Geographic Information Systems: An Introduction*, Englewood Cliffs, NJ: Prentice Hall, 1990.

Journal Articles

Olsson, L. "Integrated Resource Monitoring by Means of Remote Sensing, GIS and Spatial Modeling in Arid Environments." *Soil Use and Mangement* 5: 30-37 (1989).

Running, S.W., et. al. "Mapping Regional Forest Evapotranspiration and Photosynthesis by Coupling Satellite Data with Ecosystem Simulation." *Ecology* 70: 1090-1101 (1989).

A FUTURISTIC GIS

SOME EXAMPLES OF ADVANCED ANALYTICAL PROCEDURES

The best way to predict the future is to invent it.
—*Allan Kay*

Spatial analysis is more than mapping and spatial database management. It involves deriving new information to express relationships based on the relative positions of map features. This section establishes a framework for spatial analysis and demonstrates several of its important aspects.

Analyzing the Analytical 30
The Unique Character of Spatial Analysis

GIS mapping and management capabilities are becoming common in the workplace. The mapping revolution will be complete when most office automation packages offer GIS at the touch of a button. So can the GIS technocrats declare a victory and fade into a well-deserved (and well-appointed) retirement? Is that all there is? Or have we merely attained another milestone along GIS's evolutionary path?

GIS is often described as a "decision-support tool." Currently most of that support comes from its data mapping and management (inventory) capabilities. The abilities to geo-query datasets and generate tabular and graphical renderings of the results are already recognized; they are invaluable tools but are still basically data handling operations. It is the ability of GIS to analyze the data that will eventually revolutionize the way we deal with spatial information.

The heart of GIS analysis is the spatial/relational model, which expresses complex spatial relationships among map entities. The relationships can be stored in the data structure (topology) or derived through analysis. For example, the cascading relationships among river tributaries can be encapsulated within a dataset. The actual path a rain drop takes to the sea and the time it takes to get there, however, must be derived from the data by optimal path analysis. Derivation techniques like these form the spatial analysis toolbox.

The term "spatial analysis" has assumed various definitions over time and discipline. To some, the geo-query for all locations of dense, old Douglas fir stands from the set of all forest stands is spatial analysis. But to the GIS purist, the inquiry is a nonspatial database management operation. It involves manipulating the attribute database and producing a map as its graphical expression, but it doesn't involve spatial analysis. Spatial analysis, strictly defined, involves operations in which results depend on data locations—move the data, and the results change. For example, if you move a bunch of elk in a park their population center moves, but the average weight for an elk doesn't. That distinction identifies the two basic types of geo-referenced measures: spatially dependent or independent. The population center calculation is a spatially dependent measurement, and the average weight considering the entire population is independent. Note that the term "measurement" is a derived relationship, not a dataset characteristic. Spatial analysis involves deriving new spatial information, not repackaging existing data.

With that definition diatribe under your belt, you have one more distinction to complete the conceptual framework for spatial analysis: derivation mechanics, whether the data are spatially aggregated or disaggregated. In the elk example, the average weight for the entire population in the park is spatially independent and aggregated. Spatial patterns can be inferred, however, if disaggregated analysis is employed by partitioning space into subunits and calculating independent measures. Such analysis might reveal that the average weight for an elk is higher in one portion of the park than it is in another.

These concepts are illustrated in figure 30.1. There are 16 field samples (samples #1-16), their coordinates (X,Y) and activity values for two 24-hour periods (P1 in June and P2 in August). Note the varying levels of activity: 1 to 42 for Period 1 and 0 to 87 for Period 2. Also note that the average animal activity increased (19 to 23), as well as its variability (12 to 26). These traditional statistics tell us things have changed, but they fail to tie the changes to the ground.

A simple spatial summary of the data's geographic distribution is its centroid, calculated as the weighted average of the X and Y coordinates. That's done by multiplying each of the sample coordinates (X_i,Y_i) by the

Animal Activity Data

Sample	X	Y	P1	P2
#1	1000	1000	11	4
2	1000	1500	19	9
3	1000	2000	8	0
4	1000	2500	0	0
5	1500	1000	27	25
6	1500	1500	12	2
7	1500	2000	14	4
8	1500	2500	2	0
9	2000	1000	10	6
10	2000	1500	17	22
11	2000	2000	34	42
12	2000	2500	22	33
13	2500	1000	20	16
14	2500	1500	28	43
15	2500	2000	42	87
16	2500	2500	34	68
TOTAL			300	361
AVERAGE			19	23
STD. DEV.			12	26

Sample Locations

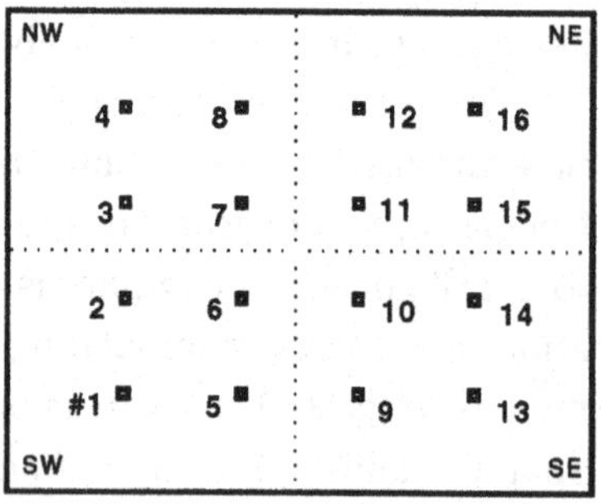

NW　　　　　　　　　　　　　　　　　　NE

4▪　8▪　　▪12　▪16

3▪　7▪　　▪11　▪15

2▪　6▪　　▪10　▪14

#1▪　5▪　　▪9　　▪13

SW　　　　　　　　　　　　　　　　　　SE

Figure 30.1. Field data of animal activity.

number of animals for a period at that location (Pi), then dividing the sum of the products by the total animal activity (SUMXiPi/SUMPi and SUMYiPi/SUMPi where i = 1 to 16). Whew!

The calculations show the centroid for Period 1 as X = 1,979 and Y = 1,728, and it shifts to X = 2,218 and Y = 1,893 for Period 2. Because the measure moved, the centroid must involve spatial analysis. Both periods show a "displaced centroid" from the geographic center of the project area. If the data were uniformly or randomly distributed, the centroid and the geographic center would align. The magnitude of the difference indicates the degree of the displacement and its direction indicates the orientation of the shift. Comparing the centroids for the two periods shows a shift toward the northeast (X = 2,218 - 1,979 = 239 meters to the east and Y = 1,893 - 1,728 = 165 meters to the north).

The centroid identifies the data's "balance point," or centrality. Another technique to characterize the data's geographic distribution is a table of the spatially disaggregated means. First the study area is partitioned into quarter-sections, or "quads." The average for the data within each quad is computed, then compared to the average of the entire area. The calculations show the following:

Period 1		**Period 2**	

NW	NE	NW	NE
6	33	1	58

SW	SE	SW	SE
17	19	4	55

Whew! So what does that tell you (other than "being digital" with maps is a pain)? It appears the southeast and northeast quads have consistently high populations (always above the period averages of 19 and 23), which squares with the centroid's northeast displacement. Also, the northeast quad consistently has the greatest overage (33 - 19 = 14 for P1 and 58 - 23 = 35 for P2), and the southwest quad has the greatest percentage decrease ([(17 - 19) - (4 - 23) / (17 - 19)] * 100 = -850%) from the averages for the two time periods.

The analysis would be even more spatially disaggregated if you were to "quad the quads," and compute their means. With this example, however, there would be only one sample point in each of the 16 subdivisions, and their mean would be meaningless. What if you "quaded the quaded quads" (8 x 8 = 64 cells)? Most of the partitions wouldn't have a sample value, so what would you do? That's where the previous discussions of spatial interpolation (Chapters 4-6) come in to fill the holes. Chapter 31 will build on the spatial interpolation surface and fill in a few of the conceptual holes as well, such as assumptions about spatial dependency, autocorrelation, and cross-correlation. Heck, by the time this topic is over, at least you'll have a useful bunch of intimidating techno-science terms to throw around.

If you're really into this stuff, consider the following calculations for the centroid and disaggregated means.

Calculations for Centroid and Disaggregated Means

	DATA				CENTROID			
					P1		P2	
Sample	X	Y	P1	P2	X_iW_i	Y_iW_i	X_iW_i	Y_iW_i
#1	1,000	1,000	11	4	11,000	11,000	4,000	4,000
2	1,000	1,500	19	9	19,000	28,500	9,000	13,500
3	1,000	2,000	8	0	8,000	16,000	0	0
4	1,000	2,500	0	0	0	0	0	0
5	1,500	1,000	27	25	40,500	27,000	37,500	25,000
6	1,500	1,500	12	2	18,000	18,000	3,000	3,000
7	1,500	2,000	14	4	21,000	28,000	6,000	8,000
8	1,500	2,500	2	0	3,000	5,000	0	0
9	2,000	1,000	10	6	20,000	10,000	12,000	6,000
10	2,000	1,500	17	22	34,000	25,500	44,000	33,000
11	2,000	2,000	34	42	68,000	68,000	84,000	84,000
12	2,000	2,500	22	33	44,000	55,000	66,000	82,500
13	2,500	1,000	20	16	50,000	20,000	40,000	16,000
14	2,500	1,500	28	43	70,000	42,000	107,500	64,500
15	2,500	2,000	42	87	105,000	84,000	217,500	174,000
16	2,500	2,500	34	68	85,000	85,000	170,000	170,000
TOTAL			300	361	596,500	518,500	800,500	683,500 TOTAL
N			16	16	300	300	361	361 SUM_{wi}
AVERAGE			19	23	X = 1,979	Y = 1,728	X = 2,218	Y = 1,893 CENTROID
STD. DEV.			12	26	X = 2,218 - 1,979 = 239		Y = 1,893 - 1,728 = 165	$SHIFT_{x,y}$

Disaggregated Means

Sample Locations

Southwest quadrant contains samples #1, 2, 5, 6.

$AVG_1 = (11 + 19 + 27 + 12) / 4 = 17$ (17 - 19 = **-2**)

$AVG_2 = (4 + 9 + 25 + 2) / 4 = 4$ (4 - 23 = **-19**)

Northwest quadrant contains samples #3, 4, 7, 8.

$AVG_1 = (8 + 0 + 14 + 2) / 4 = 6$ (6 - 19 = **-13**)

$AVG_2 = (0 + 0 + 4 + 0) / 4 = 1$ (1 - 23 = **-22**)

Southeast quadrant contains samples #9, 10, 13, 14.

$AVG_1 = (10 + 17 + 20 + 28) / 4 = 19$ (19 - 19 = **0**)

$AVG_2 = (6 + 22 + 16 + 43) / 4 = 55$ (55 - 23 = **32**)

Northeast quadrant contains samples #11, 12, 15, 16.

$AVG_1 = (34 + 22 + 42 + 34) / 4 = 33$ (33 - 19 = **14**)

$AVG_2 = (42 + 33 + 87 + 68) / 4 = 58$ (58 - 23 = **35**)

Respecting Mapped Data

Analyzing Spatial Dependecy
Within A Map

31

Chapter 30 identified two measurements that characterize the geographic distribution of field data: centroid and spatially disaggregated means. Both techniques reduce findings to discrete, numeric summaries. The centroid's X,Y coordinates identify the data's balance point, or centrality. The spatially disaggregated means are expressed in a table of localized averages for an area's successive quarter-sections. Both techniques reveal the geographic bias in a dataset, but fail to map the data's continuous distribution. That's where spatial interpolation comes in to estimate the characteristics of unsampled locations from nearby sampled ones.

Consider the 3-D plot in the center of figure 31.1. It identifies a weighted nearest-neighbors interpolated surface of the geographic distribution for Period 2 animal activity data. Note that the peak in the northeast and the dip in the northwest are consistent with the centroid and dissaggregated means characterizations discussed in Chapter 30. With the graphical rendering, however, you can "see" the subtle fluctuations in animal activity within the landscape. High activity appears as a mountain in the northeast and a smaller hill to the south—sort of a two-bumper distribution.

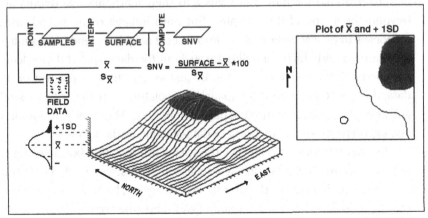

Figure 31.1. Calculation of Standard Normal Variable (SNV) map (univariate spatial analysis).

The surface looks cool and is generally consistent with the bias reported by the centroid and aggregated means, but is it really a good picture of the distribution? What are the assumptions ingrained in spatial interpolation? How well do they hold in this case? That brings us to the concept of spatial dependence—what occurs at one location depends on the characteristics of nearby locations, and near things tend to be more related than distant things. Spatial dependence can be negative (near things are less alike) or positive (near things are more similar). But common sense and most interpolation techniques are based on positive spatial dependence.

That implies a measure of spatial dependence within a dataset should provide insight into how well spatial interpolation might perform. Such a measure is termed "spatial autocorrelation." For the techy-types, autocorrelation (in a nonspatial statistics context) means residuals tend to occur as clumps of adjacent deviations on the same side of a regression line—a bunch above, then a bunch below (high autocorrelation), which is a radically different situation than every other residual alternating above then below (low autocorrelation). For the rest of us, it simply means how good one sample is at predicting a similar sample (or a near sample in GIS's case).

It shouldn't take a rocket scientist to figure out that high spatial autocorrelation in a set of sample data should yield good interpolated results. Low autocorrelation should lower your faith in the results. Two measures often are used: the Geary index and the Moran index. The Geary index compares the squared differences in value between neighboring samples with the overall variance in values among all samples. The Moran index is calculated similarly, except it's based on the product of values. The equations have lots of subscripts and summation signs and their mathematical details are beyond the scope of this chapter, but both indices relate neighboring responses to typical variations in the dataset. If neighbors tend to be similar, yet there's a fairly high variability throughout the data, spatial dependency is rampant. If the neighbors tend to be just as dissimilar as the rest of the data, there isn't much hope for spatial interpolation. Yep, this is techy stuff, and I bet you're about to turn the page. But hold on! This stuff is important if you intend to go beyond mapping or data-painting by the numbers.

In a modern GIS, you can click the spatial interpolation button and generate a map from field data in a few milliseconds. But you could be on thin ice if you simply assume it's correct—ask Geary or Moran if it's worth generating a Standard Normal Variable (SNV) map (the tremendously useful map shown on the right side of figure 31.1) to identify statistically unusual locations. The procedure calculates a normalized difference from the average for each interpolated location, effectively mapping the standard normal curve in geographic space. The planimetric plot in the figure identifies areas of unusually high animal activity (shaded blob in the extreme northeast) as locations that are one or more standard deviations greater than the average (depicted as the line balancing the surface's volume).

So what? Rather, so where! If your data show lead concentrations in the soil instead of animal activity, you can identify areas of significantly high lead levels. Or, if your data indicate lead concentration in blood samples, you can identify pockets of potentially sick people. If your data are monthly purchases by customers, you can see where the big spenders live. The SNV map directs your attention to unusual areas in space. The next step is to relate such unusual areas to other mapped variables.

But that step takes us into another arena—from univariate to multivariate spatial analysis. Univariate analysis characterizes the relationships within in a single mapped variable, such as an SNV that locates statistically unusual areas. Multivariate analysis, however, uses the coincidence among maps to build relationships among sets of mapped data. For example, figure 31.2 calculates a percent-change map between the two periods. The planimetric plot in the figure shows areas of increased activity in 10 percent contour steps. The shaded area identifies locations that increased more than 50 percent. Now if these were sales data, wouldn't you like to know where the big increases recently occurred? Even better, statistically relate these areas to other factors, such as advertising coverages, demographics, or whatever else you might try as a driving or correlated factor. But that's for Chapter 32.

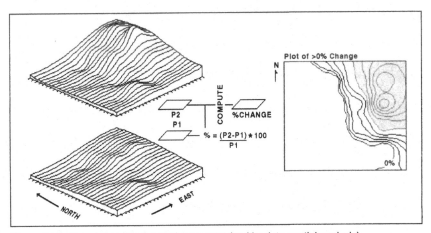

Figure 31.2. Calculation of percent change map (multivariate spatial analysis).

Tech-nacious Users

Analyzing Spatial Dependecy Between Maps

32

Most traditional mathematical and statistical procedures extend directly into spatial analysis. At one extreme, GIS is simply a convenient organizational scheme for tracking important variables. With geo-referenced data hooked to a spreadsheet or database, drab tabular reports can be displayed as colorful maps. At another level, geo-referencing serves to guide the map-ematical processing of delineated areas, such as total amount of pesticide applied in each state's watersheds. Finally, spatial relationships themselves can form the basis for extending traditional math/stat concepts.

The top portion of figure 32.1 identifies a unique spatial operation: point pattern analysis. The random pattern is used as a standard that assumes all points are located independently and are equally likely to occur anywhere. The average distance between neighboring points under the random condition is based on the density of points per unit area: $1/(2*density)^2$, to be exact. Now suppose you record the locations for a set of objects (e.g., trees), or events (e.g., robberies) to determine if

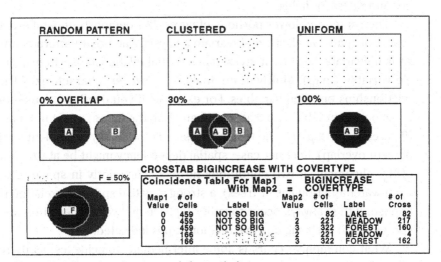

Figure 32.1. Pattern and cross-correlation analysis.

they form a random pattern. The GIS computes the actual distance between each point and its nearest neighbor, then averages the distances. If the computed average is close to the random statistic, randomness is indicated. If the computed average is smaller, a clumped pattern is evident. And if it approaches the maximum average distance possible for a given density, the pattern occurs uniformly.

An alternate approach involves a roving window, or filter. It uses disaggregated spatial analysis as it moves about the map calculating the number of points at each position. Because the window's size is constant, the number of points about each location indicates the relative frequency of point occurrence. A slight change in the algorithm generates the average distance between neighboring points and compares it to the expected distance for a random pattern of an equal number of points within the window. Whew! The result is a surface with values indicating the relative level of randomness throughout the mapped area.

The distinction between the randomness statistic for the entire area and the randomness surface is important. It highlights two dominant perspectives in statistical analysis of mapped data: spatial statistics and geostatistics. Generally, spatial statistics involves discrete space and a set of predefined objects, or entities. In contrast, geostatistics involves continuous space and a gradient of relative responses, or fields. Although the distinction isn't sharp and the terms are frequently interchanged, it generally reflects data structure preferences—vector for entities and raster for fields.

The center and lower portions of figure 32.1 demonstrate a different aspect of spatial analysis: multivariate analysis. The point pattern analysis is univariate, because it investigates spatial relationships within a single variable (map). Multivariate analysis, however, characterizes the relationships between variables. For example, if you overlay a couple of maps on a light-table, two features (A and B in the figure) might not align (0 percent overlap). Or the two features could be totally coincident (100 percent overlap). In either case, spatial dependency might be at the root of the alignment. If two conditions never occur jointly in space (e.g., open water and Douglas fir trees), a strong negative spatial correlation is implied. If they always occur together (spruce budworm infestation and spruce trees), a strong positive relationship is indicated.

What commonly occurs is some intermediate coincidence, as there would be some natural overlap even from random placement: if there are only two features and they each occupy half of the mapped area, you would expect a 50 percent random overlap. Deviation from the expected overlap indicates spatial crosscorrelation, or dependency between maps, but two conditions keep the concept of spatial crosscorrelation from being that simple. First, instances of individual map features are discrete and rarely occur often enough for smooth distribution. Also, the multitude of features on most maps results in an overwhelmingly complex table of statistics.

Recall the plot of "big increase" in animal activity (>50 percent change between Period 1 and Period 2) discussed in Chapter 31. It was a large shaded glob in the northeastern corner of the study area. I wonder if the occurrence of the big increase relates to another mapped variable, such as cover type? The crosstab table at the bottom of figure 32.1 summarizes the joint occurrences in the area of Big Increase and cover type classes of Lake, Meadow, and Forest. The last column in the table reports the number of grid cells containing both conditions identified on the table's rows. Note that Big Increase jointly occurs with Meadows only four times, and it occurs 162 times with Forests. It never occurs with Lakes. (That's fortuitous, because the animal can't swim.) A gut interpretation is that they are "dancing in the woods," as the Big Increase (I) is concentrated in the Forest (F).

But can you jump to that conclusion? The cell count has to be adjusted for the overall frequency of occurrence. The diagram to the left of the table illustrates this concept. From I's point of view, 98 percent of its occurrence is associated with F. But from F's perspective, only 50 percent of its area coincides with I. Aaaaahhhh! All this statistical mumble-jumble seems to cloud the obvious.

That's the trouble with being digital with maps. The traditional mapping community tends to see maps as sets of colorful objects, while the statistics community tends to ignore (or assume away) spatial dependency. GIS is posed to shatter that barrier. As a catalyst for communication, consider figure 32.2. It identifies several examples of statistical techniques grouped by the Descriptive/Predictive and Univariate/Multivariate dimensions of the statistician and the Discrete/Continuous dimension of the GISer. At a minimum, the figure should generate thought, discussion, and constructive dialogue about the spatial analysis revolution.

One thing is certain: GIS is more different from than it is similar to traditional mapping and data analysis. Many of the mapematical tools are direct conversions of existing scalar procedures. Sure you can take the second derivative of an elevation surface, but why would you want to? Who in their right mind would raise one map to the power of another? Is map regression valid? What about cluster analysis? How would you use such techniques? What new insights do they provide? What are the restrictive and enabling conditions?

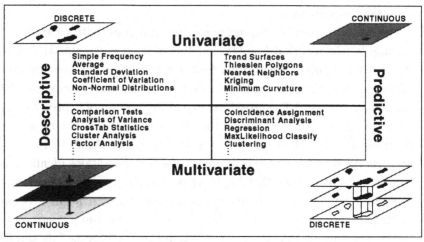

Figure 32.2. Dimensions of spatial statistics.

Another thing is certain: GIS raises as many questions as it answers. As we move beyond mapping toward spatial reasoning, the linkage between the spatial and quantitative communities will strengthen. Both perspectives will benefit from realistic physical and conceptual renderings of geographic space. But the linkage must be extended to include the user. The increasing complexity of GIS results from its realistic depiction of spatial relationships. Its descriptions of space are vivid and intuitive. Its analyses, however, can be confusing and foreign to new users. The outcome of the pending spatial analysis revolution hinges as much on users' acceptance as on technological development.

See pages 175-92 for a discussion of addtional analytical procedures and mathematical techniques.

Recommended Reading

Books

Barber, G.M. "Univariate Inferential Statistics" and "Statistical Relationships Between Two Variables." Parts 2 and 3 in *Elementry Statistics for Geographers*, New York, NY: The Guilford Press, 1988.

Berry, J.K. "Roving Windows: Assignment of Neighborhood Characteristics," "Assessing Variability, Shape and Pattern of Map Features," "Overlaying Maps and Summarizing the Results," and "Slope, Distance, and Connectivity: Their Algorithms." Topics 3, 5, 7, and 9 in *Beyond Mapping: Concepts, Algorithms, and Issues in GIS*, Fort Collins, CO: GIS World Books, 1993.

Cressie, Noel. "Statistics for Spatial Data," "Geostatistics," and "Special Topics in Statistics for Spatial Data." Chapts. 1, 2, and 5 in *Statistics for Spatial Data*, New York, NY: John Wiley & Sons, 1993.

Cromley, R.G. "Generalization: Classification and Simplification." Chapt. 5 in *Digital Cartography*, Englewood Cliffs, NJ: Prentice Hall, 1992.

Dent, B.D. "Techniques of Quantitative Thematic Mapping." Part 2 in *Cartography: Thematic Map Design,* 2d ed., Dubuque, IA: Wm. C. Brown Publishers, 1990.

Myers, D.E (ed.). "Spatial Statistics." Topic 6 in *Environmental Modeling with GIS*, ed. M.F. Goodchild, B.O. Parks, and L.T. Steyaert, Oxford, UK: Oxford University Press, 1993.

Unwin, D. "Map Comparison." Chapt. 7 in *Introductory Spatial Analysis*, New York, NY: Methuen, 1981.

Journal Articles

Bracken, I., and D. Martin. "The Generation of Spatial Population Distributions from Census Centroid Data." *Environment and Planning* A 21(8): 537-544 (1989).

Johnson, L.B. "Analyzing Spatial andTemporal Phenomena Using Geographical Information Systems: A Review of Ecological Applications." *Landscape Ecology* 4(1): 31-43 (1990).

Epilog

The Human Factor in GIS Technology

*Experience is what you get when
you don't get what you want.*

The corollary to this axiomatic truth is "Learn from others' mistakes so you won't have to make them all yourself." As GIS moved from its infancy in the early 1970s to its present maturity, the school of hard knocks coughed up an ample set of good, bad examples. We may not know what is best for all GIS environments, nor have a formula for assured success, but the growing layers of scar tissue in the GIS community indicate some of the paths not to be taken.

An Experiential GIS

Let me illustrate with an early experience in the application of GIS to land use planning. It was a class project for a graduate course in GIS at Yale University in the spring of 1980. The saga pits a naive and somewhat dim-witted assistant professor backed by a covey of bright students against an enraged portion of the populace of Guilford, Connecticut, a picturesque town on Long Island Sound. The early stages of the project were typically blissful, with energy focused on database development within the tender arms of academia. The students feverishly encoded 20 data layers for the nearly 70-square-mile town, including the usual set from standard map sheets, augmented with special town maps such as zoning, sensitive soils, and land use. That in itself was a great learning experience, given the pre-Paleolithic tools of the time.

Where we went wrong was in attempting to address a "real-world" problem. The town recently had completed its Comprehensive Plan of Development and Conservation as required by the Coastal Wetlands Act. It was the result of several years' effort among citizen groups and town officials and consisted of 21 policy statements, such as "protect inland wetlands ... from contamination and other modifications," "preserve farmlands," and "encourage development near or within existing developed areas."

Because all 21 statements had spatial components, it seemed natural to map the conceptual model embodied in the plan. Using a three-tier ranking scheme of suitable, less suitable and unsuitable, each policy statement was interpreted into a map of suitability for development. For example, farmland was identified on the town's land use map and then designated as less suitable for development. Similarly, the policy goal of protecting wetlands caused such areas on the sensitive soil map to be designated as unsuitable for development. In contrast, areas near or within existing development as shown on the land use map were identified as suitable for development. Following the plan's organization, the statements were grouped into four submodels—Water and Sewage, Growth, Preservation, and Natural Land Use—and combined into one overall suitability map.

Near the end of the term, enthusiasm was high and success seemed imminent. That was until we hosted a town meeting to present the results. Students served refreshments and proudly stood by *their* computer-generated maps draped on the high school walls. As fledgling GIS technocrats, they were eager to enlighten the audience as to the importance of the technology and the elegance of the map analysis process. However, the congregation seemed bored by the techno-babble and focused their collective attention on the suitability map. Once they located their property (you know, the parcel they were holding to pay for Sonny's college tuition), they did one of two things— (1) profusely thanked the students for an undoubtedly thorough job and promptly departed to relieve the babysitter, or (2) locked the last student in the reception line in animated debate and, once pried loose, sat down in seething hostility. In less than a half-hour we had distilled our audience to a residue of enraged citizens who held "unsuitable" property. We left about midnight and had to sneak back in the morning before basketball practice to recover what maps we could from the walls.

What went wrong? We did our homework. We developed an accurate database. We conscientiously translated *their* policy statements into maps and composited them as implied by *their* plan. We thought we had done it all, and we had, from a GIS-centric perspective. What we missed is the GIS wild-card—the human factor. The textual rendering of the comprehensive plan was comfortably innocuous because it lacked threatening spatial specificity. It was acceptable to outline a set of amorphous goals, then proceed with incremental planning whenever a developer proposed a specific parcel. If contention arose, there were always planning variances, exceptions, mitigation, and the ultimate recourse of lawyers and judges. That was the way things had always been done—the natural law of land use planning. Presenting the actual spatial ramifications of a comprehensive plan was like poking a stick into a den of rattlesnakes. As any seasoned planner knows, you plan, then move on before you implement: It's dangerous out there.

Being a slow learner and somewhat bent on self-flagellation, I decided to extend the project the following year. First, the students refined the database and the model, then determined the most limiting policy goals by systematically relaxing criteria in successive runs (sensitivity analysis). Armed with that insight, we solicited the help of the three town commissions instrumental in the plan's development: the Economic Development Commission, the Planning and Zoning Commission, and the Conservation Commission. At working meetings, policy-rating questions were posed to each group and *their* hierarchial orderings of the policy statements where used for subsequent model runs.

The results were three maps of overall suitability that expressed alternative interpretations of the plan. For example, the Conservation Commission's interpretation of "protect inland wetlands" was emphatic. Because Guilford is damp almost everywhere, 83 percent of the town was deemed unsuitable for development. The Economic Commission, however, believed sound engineering would protect wetlands: This lowered the wetland policy's rating and resulted in only 21 percent being unsuitable. By simply subtracting the two maps, the areas of agreement and contention were identified. The comparison map and the alternative interpretations by each of the commissions were published in the local paper and "healthy *a priori* discussion ensued." Most importantly, we minimized GIS student casualties.

The Guilford experience[1] altered my perspective of what GIS is (and isn't). Yes, it's hardware and software. It's a database. And GIS models. But, in actuality, it's the domain of the end user and those impacted. Neither GIS jerk nor jock can solve someone else's concern with rapid geo-query and a palette of 64,000 colors draped on a three-dimensional plot. In real-world applications, GIS acts as a communication tool for understanding important factors, their interactions, and various interpretations of both.

AN UNDERSTANDING GIS

Effective GIS applications have something to do with data and everything to do with understanding, creativity, and perspective. In the Information Age the amount of knowledge doubles every 14 months or so. With the advent of the information super highway, this periodicity will likely accelerate. But does more information translate directly into better decisions? Does the Internet enhance information exchange or overwhelm it? Does the quality of information correlate with the quantity of information? Does the rapid boil of information improve the broth of decisions?

GIS is a prime contributor to the landslide of information, as it releases terra bytes of mapped data on an unsuspecting (and seemingly ungrateful) public. From a GIS-centric perspective, we're doing a bang-up job. Lest I sound like a malcontent, let me challenge that observation. My perspective might not meet the critical eye of a good philosopher, but that's not the objective: The following thoughts simply explore the effects of rapid information transit on our changing perceptions of the world around us.

Let's split hairs on some important words borrowed from the philosophers—*data, information, knowledge,* and *wisdom.* While they are often used interchangeably, they are actually distinct from one another in some subtle and not-so-subtle ways.

First, *data,* "factoids" of the Information Age: Data are bits of information, typically, but not exclusively, in numeric form such as cardinal numbers, percentages, statistics, etc. Obviously, data are increasing at an incredible rate. Coupled with the barrage of data is a requirement for the literate citizen to have a firm understanding of averages, percentages, and, to a certain extent, statistics. More and more, these types of data dominate the media and are the primary means used to characterize public opinion, report trends, and persuade specific actions.

Second, *information,* closely related to data: The difference is that we tend to view information as more word-based and/or graphic than numeric. Information is data with explanation. Most of what is taught in school is information. Because it includes everything that is chronicled, the amount of information available to the average citizen increases substantially each day. The power of technology to link us to information is phenomeal—for proof, simply "surf" the exploding number of home pages on the World Wide Web.

The philosophers' third category is *knowledge,* information within a context: Data and information that explain a phenomenon become knowledge. Knowledge probably does not double at fast rates, but that has more to do with the learner and processing techniques than with what's available. In other words, knowledge is data and information once we can process and apply them.

The last category, *wisdom*, certainly does not double at a rapid rate. It is the application of all three previous categories with some intangible additions. Wisdom is rare and timeless and is important because it is rare and timeless. We seldom encounter new wisdom in the popular media, nor do we expect a deluge of newly derived wisdom to spring forth from our computers each time we log on.

Knowledge and wisdom, like gold, must be aggressively processed from tons of nearly worthless material. Increasing the quantities of data and information does not generate the greater amounts of the knowledge and wisdom needed to solve pressing problems. Increasing the processing "throughput" by efficiency gains and new approaches might.

So, how does that philosophical diatribe relate to GIS ? What's our role within the framework? What do we deliver—data, information, knowledge, or wisdom? We can affect all four if GIS is appropriately presented, nurtured, and applied; and if we recognize technology's role as an additional link the philosophers failed to note.

Understanding sits between information and knowledge. Understanding involves honest dialogue among various interpretations of data and information to reach common knowledge and wisdom. Note that understanding is not a thing, but a process. It's how concrete facts are translated into the slippery slope of beliefs. It involves a clash of values, tempered by judgment based on the exchange of experience. Technology, particularly GIS, has a vital role to play in that process. We need to deliver not only spatial data and information, but also a methodology for translating them into knowledge and wisdom.

In our earliest GIS encounters we viewed maps as images; automated cartography provided rapid updating and redrafting of traditional map products. We then focused on deriving and organizing mapped data and the field quickly progressed from computer mapping to spatial database management which provides efficient storage and retrieval of vast amounts of land-based data in both tabular and graphic form. From that view, GIS functions like a cash register to record land transactions. Now we see GIS as a toolbox of map analysis operations in which entire maps are treated as variables and related within a specific context. The GIS toolbox translates mapped data into spatial information.

Tomorrow's GIS will build on the cognitive basis together with spatial databases and analytical operations. Our new view will push GIS beyond data mapping, management and modeling toward spatial reasoning and a dialogue of ideas. In a sense, GIS will transform the toolbox to a sandbox, in which alternative perspectives are constructed and discussed and common knowledge and wisdom evolve.

The end users of GIS need to be fully engaged in that process, not only with the encoded and derived products of GIS. If the democratization of GIS is to go beyond graphic user interfaces and attractive icons it will require the GIS priesthood and technocrats to explain concepts in layperson's terms and render the conceptual expressions of geographic space accessible through intuitive means divorced from macro code.

I hope we consider the importance of knowledge and wisdom in the Information Age, and eagerly grasp the opportunity GIS has to contribute to their derivation. I fear that GIS "factlets" masquerading as knowledge in the Information Age will obscure the importance of wisdom. I fear that our all-consuming focus on maps and data will distract us from assimilating the significance embedded in spatial information and communicating the ideas it spawns. GIS presents an opportunity to empower people with new decision-making tools as well as the danger of trapping them in new technology and avalanches of data. What we have accomplished is necessary, but not yet sufficient, for effective GIS solutions.

Like the automobile and indoor plumbing, GIS won't be an important technology until it fades into the fabric of society and is taken for granted. It must become second nature for accessing information *and* translating it into knowledge. We must shift its emphasis beyond mapping to spatial reasoning.

1. See Berry and Berry, "Assessing Spatial Impacts of Land Use Plans," 1988, in *Journal of Environmental Management*, 27: 1-9; and Berry, et. al., "Analysis of Spatial Ramifications of the Comprehensive Plan of a Small Town," 1981, in *Proceedings*, 41st Symposium, American Congress of Surveying and Mapping.

The Most Beautiful Formulae in GIS

By Nigel Waters

One of the most important characteristics of a GIS is that it can be used as a spatial decision support system (SDSS). A GIS can act like an SDSS because it is possible to subject the data stored within a GIS to mathematical manipulations. According to Gail Langran (1989), the ability to analyze data distinguishes first-generation GISs (which could do no more than store, retrieve, and display data) from the second-generation systems that have these powerful processing capabilities.

These mathematical manipulations require a variety of equations and procedures that have their own simplicity, logic, and beauty. The purpose of this appendix is to introduce you to some of the key areas of mathematical manipulation used in a GIS. The appendix stresses the elegance of these formulae and equations, as well as their cleverness and usefulness—a common approach in many expository, mathematical texts (see, for example, Salem et al.,1992; Gardner, 1978; and Dunham, 1991). The idea is that if you have a deep, intuitive understanding of the logic and elegance of the equations then you'll more likely remember and retain that knowledge. In addition, for each formula one or more key references are included that allow the reader to dig much deeper into this treasure trove of powerful mathematical tools.

To be included, these formulae meet some of the following criteria:

- They are widely used.

- They are the fundamental building blocks of many key GIS algorithms and analytical procedures.

- They impart powerful ideas (see Papert, 1980, for an extended discussion of this concept).

- They should be representative of and related to other key formulae.

Most vector-based GISs traditionally used a hybrid structure for data storage (Healey, 1991). That means locational information is stored separately from attribute data. Attribute data represent the characteristics of whatever is being mapped. Thus, many GISs, such as ARC/INFO from Environmental Systems Research Institute, Inc. (ESRI), Redlands, Calif. (see GIS World

Sourcebook 1995 from GIS World, Inc., Fort Collins, Colo., for a detailed description of GIS packages and the companies that market them), use proprietary software to store the spatial information and a relational database for the attribute information. The proprietary spatial software may be their own, as in the case of ESRI, or commercially available computer-aided design (CAD) software. Occasionally some systems use integrated approaches, but that is much less common. I follow the hybrid approach here and divide the algorithms and equations into two groups: spatial and nonspatial.

SPATIAL FORMULAE

1. The Oldest of Them All

Perhaps the most beautiful of all is the oldest. It is the formula for distance between two points and is based on Pythagoras' theorem from the sixth century B.C. (see Dunham, 1991, for historical details).

$$D = (\mid X_1 - X_2 \mid ^2 + \mid Y_1 - Y_2 \mid ^2)^{1/2}$$

The two points are represented by their X and Y coordinates in a Cartesian space. Thus, X_1 and Y_1 represent the coordinates of the first point, and X_2 and Y_2 represent the coordinates of the second point. D is the straight line distance between the two points. The vertical bars mean take the absolute value. The 1/2 exponent is, of course, the square root.

The equation is a fundamental building block in many GIS algorithms. Distance is important in many GIS procedures. For example, if a GIS analyst uses a spatial interpolation method, such as a distance weighted moving average (see spatial formula No. 7), to produce an interpolated surface or "drape," he or she probably will interpolate a square grid of points from an irregularly spaced dataset. Each point on the grid will be a distance weighted average of the six nearest points in the original dataset. The above equation is required to determine the six nearest points.

Many other algorithms are made more efficient if the number of possibilities that have to be considered are reduced, and that is achieved often by considering only the closest points. The Very Important Points Algorithm for choosing points to be part of a triangulated irregular network (TIN) model requires distance calculations. The TIN model is a more efficient alternative to the digital elevation model (DEM) for representing a surface (see Unit 39 in the Core Curriculum developed by the National Center for Geographic Information and Analysis (NCGIA) by Goodchild and Kemp, 1990; and also Weibel and Heller, 1991).

If the "twos" in the exponents are replaced by R, then R is known as the Minkowski metric. When R=1 we have the so-called city block or Manhattan distance. That formula has been used ingeniously in location-allocation modeling where the analyst uses the GIS to suggest the optimum location for a facility, such as a fire station or a fast food restaurant, and then determines the allocation of residences or potential customers to that facility. The facility users are allocated to the facility to which they live closest, hence the need for distance calculations. Location-allocation modeling is one of the most useful applications of GIS in the public and private sectors. Excellent reviews are provided by Ghosh and Rushton (1987). It is interesting that ESRI has seen fit to include state-of-the-art location-allocation modeling procedures only in the most recent revision of ARC/INFO (Version 7). However, the procedure has been incorporated into Caliper Corp.'s TransCAD package for several years.

Distance calculations are important in clustering algorithms, Voronoi or Thiessen polygon algorithms, and nearest neighbor routines in point pattern analysis that are used to determine the adequacy of spatial sampling procedures.

2. The Intersection of Two Lines

$$X_i = - (a_1 - a_2) / (b_1 - b_2)$$
$$Y_i = a_1 + b_1 X_i$$

In the equations above, the b terms represent the slope of the first and second lines, and the a terms represent the intercept with the Y or vertical axis in the Cartesian coordinate system. The equations for the two straight lines from which these terms are derived are given by the following equations:

$$Y_1 = a_1 + b_1 X_1$$
$$Y_2 = a_2 + b_2 X_2$$

The initial formulae, which provide the X and Y coordinates of the point of intersection (point X_i, Y_i), are another beauty from the field of coordinate geometry, or as GIS techies say, "COGO." Coordinate geometry is one of the most useful branches of mathematics for GIS design. It is also one of the more accessible, because the rudiments of the subject usually are taught in high school mathematics classes. There are many useful texts in this area. One of the more technical and comprehensive is Hartley (1960).

A related area of mathematics that is equally important is computational geometry, which is concerned with the development of algorithms and procedures that will provide solutions to geometric problems quickly and efficiently on computers. There is now an annual symposium on computational geometry sponsored by the Association for Computing Machinery (ACM) and the special interest groups for Graphics, Automata and Computability Theory of that organization. An advanced text on computational geometry is by Preparata and Shamos (1985). The authors begin their book by noting that the original motivation for geometry among the Greeks and Egyptians was the need to tax lands accurately and fairly and to construct buildings. Of course, these are still important motivations behind today's cadastral and land-related GISs. Algorithm efficiency and speed are vital in GIS, because the operation being considered may have to be performed on tens or hundreds of thousands of lines or polygons. An inefficient algorithm may slow down even the fastest computer.

The formulae for the intersection of two lines is the cornerstone of many point-in-polygon and polygon overlay routines. And where would GIS be without them? Point-in-polygon routines are extremely important in GIS operations. Frequently we need to know how many observations fall inside a polygon and how many fall outside. The point-in-polygon operation is straightforward: a line is drawn vertically upward from the point in question. The number of times that line crosses the boundary of the polygon being considered is recorded. If that number is even, the point is outside the polygon; if it is odd, the point must be inside. The algorithm is surprisingly robust and works for polygons that have weird and wonderful shapes, but there are many "special cases" (discussed below) requiring careful consideration.

The polygon overlay operation involves finding the intersecting arcs of two different coverages (Unit 34 of the NCGIA Core Curriculum). Overlay operations are required for combining the properties of two or more coverages and, thus, have been important in environmental impact studies since the days when such operations were performed mechanically and manually. Thus, Ian McHarg (1969) used photographic overlay procedures to determine the ideal location for transportation corridors, among other things, in his classic text Design with Nature. Polygon overlay is important for buffering and windowing operations in a vector-based GIS. Buffering operations allow the user to put a small polygon around a point, line, or area. For example, a buffer might be placed around a river or a lake to protect such features from environmental damage that might result from logging. A window might be used to take a look at what lies within the window. Thus, the GIS user could find out the population characteristics of the people within the window. Alternatively the window might be used as a moving spatial averaging or filtering device to generate a smoother surface.

Before we leave this topic we should note that the above formulae solve the intersection problem, but only for simple situations. In the real world there are a host of so-called "special cases," which arise with depressing frequency. Thus, the formulae above assume lines of infinite length when we really only want to see if the lines cross somewhere on our map. Vertical lines that have an infinite slope and horizontal lines that have no slope may also cause complications. These and other difficulties are addressed in an article by David Douglas that has been reprinted several times (see, for example, Douglas, 1990).

3. The Gravity Model

$$I_{ij} = k \, ((P_i) \, (P_j) / (d_{ij})^2)$$

The gravity model was used to predict traffic flows in early transportation models from the late 1950s. It is developed extensively in the writings of William Warntz and his students who pioneered the model (see Coffey, 1988, for a review of this work). It states that the amount of interaction, I, between two places, i and j, is equal to some constant, k, multiplied by their population product divided by the distance between them squared. The equation has been used in various forms. Often the population terms are raised to a power, which is determined empirically using least squares regression procedures. The exponent on distance may also be fitted empirically. A full review of these types of models and their applications in the literature may be found in Taylor (1975). The Pi and Pj terms may be used to represent other variables besides population that have an influence on interaction, and the distance term too may represent such surrogates for distance as time and cost.

The model can be extended and manipulated to produce population potential models important in marketing and popularized in GIS packages such as SPANS, now owned by PCI, Inc., Richmond Hill, Ontario, Canada. The SPANS package has been used by such firms as Miracle Mart, Ontario, to advise on new supermarket locations. In such studies, areas of high population potential are important indicators of where to place a new store.

4. The Entropy Maximizing Model

$$T_{ij} = A_i \, O_i \, B_j \, D_j / e^{(b \, c_{ij})}$$

The equation above, in which T_{ij} is the expected or most likely distribution of trips between transportation zone i and transportation zone j, was the basis of the entropy maximizing models developed in the late 1960s. The models provided statistical respectability for the gravity model. O_i and D_j are the number of workers in the origin zone and number of jobs in the destination zone, respectively; A_i and B_j are weights; e is the irrational number 2.718…; b measures the friction of distance; and c is the cost of travel. Entropy maximizing models have appeared in a variety of forms, including the so-called singly constrained shopping models and the doubly constrained journey-to-work models. They are discussed in great detail in Wilson and Kirkby (1980). In fact, Wilson first introduced these models to geographers, planners, and GIS analysts. A gentler introduction is provided by Gould (1984).

5. Projection Formulae

$$X = R\,(l - l_0)$$
$$Y = R \ln \tan (p/4 + f/2)$$

Fifth on the list of spatial formulae and those that explicitly incorporate distance are formulae for converting spherical coordinates from Earth's surface into the two-dimensional X and Y coordinates of the Cartesian plane. There are many of these formulae—one or more formulae for each of the dozens of map projections now commonly used in GIS. In the equations above, l and f are longitude and latitude, respectively, and X and Y are the standard Cartesian coordinates. R is the radius of the sphere, and l_0 is the central meridian with all angles measured in radians. Finally, p is the mathematical number pi. This elegant formula (from Snyder, 1987) for the oldest of all standard map projections, Mercator's, has been used to produce maps of the planetary surface of most of Earth's nearest neighbors (Snyder, 1987).

Many of the large, expensive GISs allow users to convert to Cartesian coordinates from any of a large number of projections (in addition to Snyder, 1987, formulae for these projections are discussed in Richardus and Adler, 1972, and Pearson, 1990). There are some excellent standalone packages that carry out vast numbers of these transformations. Perhaps one of the most efficient and effective of these packages is The Geographic Calculator from Blue Marble Geographics, Gardiner, Maine (Waters, 1994a).

6. Trend Surface Models

The formula for a straight line in Cartesian space is also the formula for the common regression line, and that regression equation can be extended easily into two independent dimensions to produce an interpolated trend across space:

$$Y = b_0 + b_1 X_1 + b_2 X_2 + e$$

The equation above is a first-order polynomial with two independent variables. Notation for this model varies, but here I follow that given in Davis (1973, 1986). In the equation, Y is the dependent variable whose value we wish to predict. X_1 and X_2 are the independent variables, and in this model they represent location. The final term, e, is the error or residual term. If the dependent variable is elevation (although it does not have to be) and the two independent variables are eastings and northings, respectively, our trend surface, regression equation provides a linear, interpolated surface portraying the trend in elevation.

From this simple surface we can easily move to higher order surfaces that allow us to model more complex relief. The equation as presented only has the X_1 and X_2 terms raised to their first power, so it produces a surface with no inflections—it is a plane that may or may not be tilted. If we add higher order powers of X_1 and X_2 and terms which represent combinations of their cross-products (e.g., X_1X_2 and $X_1^2 X_2$), we can then create more complex models of our interpolated surface. Indeed, the number of inflexions or bends in the surface is always one less than the highest order of the X_1 and X_2 powers we use in the equation. Complete details of all of these equations and the mathematics behind them, as well as computer programs to carry out the computations, are provided in Davis (1973) and Mather (1976).

To the above equation we can also add another independent variable, X_3, and plunge into the third dimension. So this is a powerful formula indeed. Three-dimensional trend surfaces are difficult to visualize and impossible to draw on a two-dimensional piece of paper or on a flat computer screen. But we can draw them one slice at a time or use some of the visualization techniques discussed in Rosenblum, et al. (1994) to get an impression of how a variable changes over three dimensions.

The error term or residual is important and allows us to see how good our interpolated surface is. Thus, we can use the GIS to produce a map of our model, the trend surface, and we also can use it to produce a map of the residuals or errors to show where the model fails to produce a good fit with reality.

With just one independent variable in the above equation we can model how our dependent variable, for example economic rent, changes over distance, and we have the classic bid rent line—Von Thunen's model (Wilson and Kirkby, 1980).

7. Distance Weighted Averages

Distance weighted averages are like spatial moving averages. They are important in surface interpolation routines and in attempts to smooth a surface and to increase the signal-to-noise ratio of whatever is being portrayed. The equation is simple and straightforward in its unadorned version, but the beauty of the formula is that it is possible to include all sorts of ad hoc adjustments.

$$Y_k = \Sigma\,(Y_i/D_{ik})\,/\,\Sigma\,(1/D_{ik})$$

This equation simply states that the estimated value at Y_k is given by taking each of the i points and dividing by the distance between each of the i points and point k, which is the point being estimated. These values are summed for however many points are used (4, 6, and 8 points are common). The sum then is divided by the sum of the reciprocals of all the distances. Chapter 3 in Harbaugh, Doveton and Davis (1977) discusses the effect of using different numbers of points, different distance weighting procedures and other variations on this generic version of the equation. The equation has been used extensively in the SURFACE II and III series of spatial interpolation programs (Sampson, 1978, 1994).

8. Affine and Curvilinear Transformations

Closely related to the ideas expressed in spatial formula No. 5 above and to the equation used in spatial formula No. 6 above is the idea of an affine transformation. Transformations are required when we wish to register different sets of coordinates for information from the same mapped area, but the coordinates come from a variety of maps that use different projections or coordinate grids. Such transformations come in two types: affine transformations that keep parallel lines parallel and curvilinear transformations in which that's not necessarily the case.

Affine transformations include translations (where the origin of the system is moved), scalings (where the scale of the map is changed), rotations (where the map is rotated around the origin), and reflections (where a mirror image of the map is produced). These transformations are the bread-and-butter of a host of algorithms in computer graphics. Foley and Van Dam (1984) provide an exhaustive account together with computer pseudo-code for these operations. A gentler introduction, along with BASIC programs and code, is given by Myers (1982). The four operations are defined in the following equations:

Translation:	*Scaling:*
$U = X - a$	$U = c\,X$
$V = Y - b$	$V = d\,Y$

Rotation:	*Reflection (about the X axis):*
$U = X \cos(\text{alpha}) + Y \sin(\text{alpha})$	$U = X$
$V = -X \sin(\text{alpha}) + Y \cos(\text{alpha})$	$V = f - Y$

Where U and V are the new transformed coordinates
X and Y are the original coordinates
a and b represent the number of units the Y and X axes are shifted
c and d represent the scale change for the X and Y coordinates
alpha is the angle of rotation measured anticlockwise
and f is the maximum value on the Y axis.

The transformations have several specific applications. For example, a reflection is needed to convert from a Cartesian coordinate system (in which the origin of the system is in the bottom left-hand corner and Y increases as we move up the map) to a raster display system, such as a line printer or a video monitor (in which the origin is in the top left-hand corner and Y increases as we move down the map). A series of these transformations (a translation, two rotations, and a reflection, respectively) also are needed to transform three-dimensional world coordinates into two-dimensional screen or map coordinates (see Myers, 1982).

One method that combines these transformations is to use two multiple regression equations of the following form:

$$U = b_1 + b_2 X + b_3 Y$$
$$V = b_4 + b_5 X + b_6 Y$$

These equations are of the same form as the first-order polynomial used in the trend surface. U and V and X and Y are defined as before, but b_1-b_6 are constants and coefficients that could be fitted using a multiple regression program, one run for each equation. Thus, a GIS analyst would have a series of X and Y coordinates and enter them as independent variables into the multiple regression program. In the first run the U coordinates from the second map would represent the dependent variable. Thus, the multiple regression program would yield the b_1, b_2, and b_3 coefficients for the first equation. A second run of the program using the V coordinates from the second map would allow the remaining b_4, b_5, and b6 coefficients to be found. Advanced treatments of affine geometry may be found in many textbooks (see, for example, Snapper and Troyer, 1971).

Curvilinear transformations involve the use of higher order polynomials such as the following:

$$U = b_1 + b_2X + b_3Y + b_4X^2 + b_5Y^2 + b_6XY$$
$$V = b_7 + b_8X + b_9Y + b_{10}X^2 + b_{11}Y^2 + b_{12}XY$$

This is a second-order polynomial, but more complex models might be used. Again a good discussion is found in Davis (1986). The whole topic of transformations is discussed in Goodchild (1984).

9. Estimating Slope and Aspect

A local trend surface equation can be calculated using a 2 by 2 or 3 by 3 window in a digital elevation model (DEM) from a raster GIS. Once the equation has been determined (either by the usual least squares method or using the simplified formulae found in Unit 38 of the NCGIA Core Curriculum) and put in the form shown in spatial formula No. 6 above, we can calculate the landscape's useful properties, such as slope and aspect. Using the same notation as in the trend surface equation above, these properties would be found by the following equations:

$$\text{slope} = (b_1^2 + b_2^2)^{1/2}$$
$$\text{aspect} = \tan^{-1} b_2/b_1$$

The first equation states that slope is equal to the square root of the sum of the squares of the b_1 and b_2 coefficients from the trend surface equation. These properties represent just two from a large number of geomorphometric properties that can be calculated from the DEM. The best discussion of this is found in articles by Evans (1990) and Pike (1988).

10. Fractals and Determining the Fractal Dimension

Fractals stem from a branch of mathematics that was popularized by Mandelbrot (1977, 1982). A popular treatment of Mandelbrot and his work is provided by Gleick (1987). Fractals have many uses, including shape measurements, evaluating the degree of convolution of two-dimensional shapes (hence they help with error measurements), evaluating the roughness of surfaces, and other generalization such as line operations. They can be used for data compression, and Barnsley (Chapter 3, 1988) discusses their relationship to affine transformations (see spatial formula No. 8 above). Geographical treatments are contained in Goodchild and Mark (1987) and Batty and Longley (1994).

Here I provide a simple equation for determining the fractal dimension of a line. Other methods are discussed in Goodchild and Mark (1987).

$$D = \log (n_2/n_1) / (\log (s_1/s_2)$$

D is the fractal dimension; s_1 is the step size of a pair of dividers used to measure the line the first time; s_2 is the step size of a pair of dividers used to measure the line the second time; n_1 is the number of steps used in the first measuring attempt; and n_2 is the number of steps used in the second measuring attempt.

The formula relies on the fact that the line will appear to become longer as the step size is decreased (equivalent to an increase in the scale of the map), because now we use a more accurate measuring device. The length increase is measured by the fractal dimension.

ATTRIBUTE-RELATED FORMULAE

1. The Normal Distribution or Bell Curve

The normal distribution describes many types of objects, including the characteristics of people, naturally occurring phenomena, and errors. The normal distribution is based on the principle that extreme values are rare. Thus, short and tall people are uncommon, whereas people of an average height are common. The bottom line is that most of us are pretty average. The same is true of animal and plants. And it is true of unbiased and nonsystematic errors as well. As a result, small errors, for example in digitizing, tend to be more common than large errors, which is why we use statistics to describe our errors.

For example, the root mean square is another name for the standard deviation, and in a normal distribution 68 percent of the values will lie between the value of the mean (or average) minus one standard deviation and the value of the mean plus one standard deviation. So by quoting the root mean square for our errors on an ARC/INFO tic coverage, for example, we can get a good idea of how serious the errors are.

The normal distribution also has been shown to provide a good model for the rate at which people and organizations adopt innovations. That idea figures prominently in the work of Rogers (1962). The point is that initially it is difficult to persuade individuals to adopt something new; because no one wants to take the risk. But there are a few people who are true innovators. They adopt right away; Then comes a larger group, the early adopters, followed by yet a larger group, the so-called early majority. These groups are followed by the late majority and finally the laggards, the latter being a small group of die-hards who need to be convinced before they adopt. The groups are divided on the basis of the standard deviation and the normal distribution.

To illustrate, suppose it takes the average person 12 years to adopt and the standard deviation of the distribution is four years. The innovators will adopt before four years have elapsed since the product's introduction. The early adopters will adopt between four years and eight years, and the early majority will adopt between eight and 12 years after the introduction. The late majority adopt between 12 and 16 years after introduction. Finally, the laggards adopt after the product has been on the market 16 years. Because we can use mathematical tables to determine the percentage between the mean and one or more standard deviations from the mean, we know exactly how many people fall into the five categories: 2.5 percent, 13.5 percent, 34 percent, 34 percent, and 16 percent, respectively. Now what does that have to do with GIS? Jeffrey Lane and David Hartgen (1989) used the model to illustrate the rate at which state transportation departments adopt GIS and related technology.

2. Standardizing Data

$$Z_i = (X_i - \text{Mean})/\text{Standard Deviation}$$

This simple formula allows us to standardize the values of different variables. Thus, variables that have different variances because they are measured in different units are reduced to the same measuring stick. The formula is important in cluster analysis in which a measure of dissimilarity is based on Euclidean distances (see spatial formula No. 1). If we did not standardize the variables, those with larger variances would exert a greater influence on the clustering process. The formula is also useful when our variables are distributed normally, because the formula will convert all our values to standard normal deviates and then we can use tables of standard normal deviates to determine the probability of finding values between certain ranges, among other applications. Many test statistics follow a normal distribution, so if we convert an observation to one of these test statistics and we know the mean and standard deviation of the sampling distribution of the test statistic, we can again convert it to a standard normal deviate and work out the probability of observing that particular value. That is the basis of many inferential statistical tests (see Siegel and Castellan, 1988, for a more detailed discussion).

3. The Chi-Square Statistic

The chi-square statistic is used to compare observed with expected distributions. It is ideal as a one-sample, goodness-of-fit test, because we can use it to see how our data are distributed. Thus, we can tell if a variable follows a normal or a poisson or some other distribution and, in the first two instances, that will allow us to take advantage of all the benefits described under attribute formulas No. 1 and No. 8. The formula for chi-square is as follows:

$$\chi^2 = \Sigma \left[(f_o - f_e)^2 / f_e \right]$$

Here f_o is the observed frequency, and f_e is the expected frequency. A thorough discussion of this statistic is given in Williams (1984). Another useful introduction is provided by Siegel and Castellan (1988). The use of formal hypothesis testing, and the scientific method is an approach that has been recommended by Waters (1994b) and Wellar and Wilson (1993).

4. The Kappa Statistic

Remote sensing and GIS often are lumped together. Remote sensing is a cheap, effective way to acquire large amounts of data over extensive parts of the world covered by harsh or inaccessible terrain. The journal Photogrammetric Engineering and Remote Sensing has a special section of each issue devoted to GIS and a complete issue devoted to GIS annually. The fact that remotely sensed data arrive as a grid of pixels makes them an ideal data source for a raster GIS.

One of the most common activities in remote sensing is to classify the pixels into groups that represent, for example, vegetation classes. Then we can compare the classified pixels to the vegetation that actually occurs on the ground in a number of test sites. That information usually is represented in a matrix in which each row of the matrix represents the true classes and the columns represent the predicted classes that result from the use of a procedure such as a discriminant equation (see attribute formula No. 4). In the matrix the diagonal elements represent a successful classification, i.e., where the predicted value is the same as the true value. Off-diagonal elements indicate a failure of the model. The GIS analyst wants to know how good the model is, how much better than a random allocation of pixels to the elements in the matrix. Even a random allocation would have some success. So to see how much better the model is we use the kappa statistic:

$$\text{kappa} = (d - q) / (N - q)$$

Here d is the number of cases in the diagonal cells, N is the total number of cases recorded in the matrix, and q is the number of cases expected in the diagonal cells (for a given matrix, q can be found by taking each row total multiplying by each column total and then dividing by N, an operation carried out for each row, and then all the row values are summed).

Alternative formulae and associated significance tests and references for this useful, elegant statistic can be found in Siegel and Castellan (1988). One of the most important properties to know about these statistics is their range. Ideally a statistic such as kappa should have the same range for matrices of any size, and it simplifies things if this range is from 0 to 1. Kappa conforms to these norms. A matrix that shows a classification which is no better than chance will have a kappa value of 0, while a matrix with a perfect classification in which all the entries are on the diagonal will have a value of 1, because d and N will be the same number.

5. Color Matching Systems

Color is important in GIS. It assists in visualization (Rosenblum, et al., 1994, and spatial formula No. 6), and almost every GIS text includes the obligatory section of color images—usually increasing the cost of the book by $5 to $10! A detailed introduction to the use of color and computer systems is provided by Durrett (1987) and Waters (1989). One of the problems with the use of color is that there are numerous ways of describing color. Usually a three-coordinate system is used. Jain (1989) provides many formulae for moving from one coordinate system to another. These formulae are usually matrix equations of the following format:

$$[O] = [TM] [I]$$

Here **[O]** is the 3-by-1 output vector of three coordinates describing the characteristics of the color, **[I]** is the 3-by-1 input vector describing the color characteristics in the original system, and **[TM]** is a 3-by-3 transformation matrix. A specific transformation matrix is required for each type of transformation.

6. Optimization Models

Maximize $Z = a X_1 + b X_2$ subject to constraints

Here we have the basis for all the programming models in their various guises, including linear, integer, and dynamic programming, and the so-called transportation models. They are also part of most location-allocation models, which now are incorporated into many of the most important GISs (see discussion of spatial formula No. 1). All of these models are important if we wish to use our GIS as an SDSS. Extensive discussion of optimization models is provided in Wilson and Kirkby (1980). See Densham (1991) for an SDSS discussion.

7. The Rank Size Rule

$$P_i = P_1 / i$$

This is a powerful formula with all sorts of statistical implications and applications. A hint of the treasure trove awaiting the investigation of this formula is given by Simon (1991). Basically, the formula says big things are less common than little things, and it describes log-normal distributions that are common in the data we store in GISs. The formula has been used extensively in urban studies in which the size distribution of urban centers is expected to follow this distribution, and it also has been used in hydrocarbon exploration studies in which hydrocarbon reservoirs also are expected to follow the distribution. In the former version of the model, P_i would represent the population of the town at rank i in the hierarchy, and P_1 is the population of the largest town in the system. Thus, the formula states that the second town in the urban system is half as big as the first, the third town is a third the size of the first, etc.

8. The Poisson Distribution

$$p(n) = (e^{-d})(d^n) / n!$$

Here d represents the density of the process, and $p(n)$ is the probability of observing n points in a given interval of either time or space. The formula has been used to evaluate the distribution of samples of points to see if the GIS provides adequate coverage when the data are spaced irregularly. It also can be used to generate data for traffic simulation models in which vehicles arrive independently of each other, as well as many other situations.

9. Classification and the F Statistic

$$F = \text{Within class variance} / \text{Between class variance}$$

So simple, so elegant, and yet it is the basis of most attempts to classify data in a GIS. And classification is so fundamental to GIS (Davis, 1986) use in resource management using remotely sensed data. Classification raises interesting philosophical and methodological issues. The former are detailed in Lakoff (1987), and the latter are addressed in Mather (1976) and Everitt (1993). Procedures that depend on this statistic include cluster analysis, trend surface analysis, regression analysis, and discriminant analysis.

10. The Discriminant Equation

Discriminant analysis is used extensively in remote sensing and GIS. For example, an analyst might classify pixels from a digital image into vegetation classes using discriminant functions. Although part of the general linear model, which assumes the data are distributed normally, the technique is relatively robust for departures from such assumptions and is used frequently. Excellent accounts of its application are found in Mather (1976), Davis (1986), and Tabachnik and Fidell (1989). The equation is similar to a multiple regression equation and is given by the following formula:

$$DS = b_0 + b_1 X_1 + b_2 X_2 + b_n X_n$$

Here DS is the discriminant score, b0 is a constant, and b1 to bn are coefficients that are fitted empirically according to some criterion, such as maximizing the minimum difference among classes. The equation can be written in a standardized form in which case the constant term drops out and the b1 to bn terms are replaced by standardized coefficients that indicate the relative importance of the variables in contributing to the discriminant function.

Conclusion

The formulae chosen here were selected for their elegance, simplicity, and generality, as well as their wide applicability and power. They represent a host of approaches and analytical procedures that have proved fruitful in constructing GISs and in subsequent GIS analysis. There are many others, but these few provide a glimpse into the power of a GIS as a tool for spatial reasoning and spatial decision support.

Acknowledgments

This appendix originated in a column that appeared in *GIS WORLD's* November 1994 issue. That original, much abbreviated treatment asked readers for their input concerning what they felt to be formulae and equations deserving of inclusion in a more lengthy treatment. Many readers responded to me by E-mail, and I am truly grateful to them for their suggestions. These readers included Lee De Cola, U.S. Geological Survey, who suggested the Normal or Gaussian distribution. I am still collecting ideas. If you have suggestions, please E-mail them to me at *nwaters@acs.ucalgary.ca*.

References

Barnsley, M. 1988. *Fractals Everywhere*. Academic Press, New York.

Batty, M. and Longley, P. 1994. *Fractal Cities*. Academic Press, New York.

Coffey, W. J., ed. 1988. *Geographical Systems and Systems of Geography: Essays in Honor of William Warntz*. Dept. of Geography, University of Western Ontario, London, Ontario.

Dunham, W. 1991. *Journey Through Genius: The Great Theorems of Mathematics*. Penguin Books Ltd., Harmondsworth, England.

Davis, J. C. 1973. *Statistics and Data Analysis in Geology* (First Edition). John Wiley, New York. (This edition of the book contains Fortran code for carrying out spatial and statistical operations of interest to the GIS analyst.)

Davis, J. C. 1986. *Statistics and Data Analysis in Geology* (Second Edition). John Wiley, New York.

Densham, P. J. 1991. "Spatial Decision Support Systems," Chapter 26, pp. 403-412 in Maguire et al., op. cit.

Douglas, D. 1990. "It Makes Me So Cross." Chapter 21, pp. 303-307 in Peuquet and Marble, op. cit.

Evans, I. S. 1990. "General Geomorphometry." Pp. 44-56 in A. Goudie (ed.) *Geomorphological Techniques*, Unwin Hyman, London.

Everitt, B. S. 1993. *Cluster Analysis* (Third Edition). Arnold, London.

Foley, J. and Van Dam, A. 1984. *Fundamentals of Interactive Computer Graphics*. Addison-Wesley, Reading, Mass.

Gardner, M. 1978. *Aha! Insight*. Scientific American Inc., W.H. Freeman and Company, San Francisco.

Ghosh, A. and Rushton, G. 1987. *Spatial Analysis and Location-Allocation Models*. Van Nostrand Reinhold, New York.

Gleick, J. 1987. *Chaos: Making a New Science*. Penguin, Harmondsworth, England.

Goodchild, M. F. 1984. "Geocoding and Geosampling." Pp. 33-53 in *Spatial Statistics and Models*, G. L. Gaile and C. J. Willmott, eds., Reidel Publishing Co., Dordrecht, Holland.

Goodchild, M. F. and Kemp, K. K. eds. 1990. *NCGIA Core Curriculum*. National Center for Geographic information and Analysis, University of California, Santa Barbara.

Goodchild, M. F. and Mark, D. M. 1987. "The Fractal Nature of Geographic Phenomena." *Annals*, Association of American Geographers, vol. 77, pp. 265-278.

Gould, P. 1984. *The Geographer at Work*. Routledge, New York.

Harbaugh, J. W., Doveton, J. H. and Davis, J. C. 1977. *Probability Methods in Oil Exploration*. John Wiley, New York.

Hartley, E. M. 1960. *Cartesian Geometry of the Plane*. Cambridge University Press, Cambridge, England.

Healey, R. G. 1991. "Database Management Systems." Chapter 18, pp. 251-267 in Maguire et al., op. cit.

Jain, A. K. 1989. *Fundamentals of Digital Image Processing*. Prentice Hall, Englewood Cliffs, New Jersey.

Lakoff, G. 1987. *Women, Fire and Dangerous Things*. The University of Chicago Press, Chicago.

Lane, J. S. and Hartgen, D. T. 1989. "Factors Affecting the Adoption of Information Systems in State Departments of Transportation." Paper 18, *Transportation Publication Series*, University of North Carolina at Charlotte, Charlotte, N.C.

Langran, G. 1989. "A Review of Temporal Database Research and Its Uses in GIS Applications." *International Journal of Geographical Information Systems*, vol. 3, 215-232.

Maguire, D. J., Goodchild, M. F. and Rhind, D. W. 1991. *Geographical Information Systems*. Longman Scientific and Technical, London.

Mandelbrot, B. M. 1977. *Fractals: Form, Chance and Dimension*. W. H. Freeman and Co., San Francisco.

Mandelbrot, B. M. 1982. *The Fractal Geometry of Nature*. W. H. Freeman and Co., San Francisco.

Mather, P. M. 1976. *Computational Methods for Multivariate Analysis in Physical Geography*. John Wiley, New York.

McHarg, I. L. 1969. *Design with Nature*. Doubleday, New York.

Myers, R. E. 1982. *Microcomputer Graphics*. Addison-Wesley, Reading, Mass.

Papert, S. 1980. *Mindstorms: Children, Computers and Powerful Ideas*. Basic Books, New York.

Pearson, F. 1990. *Map Projections: Theory and Applications*. CRC Press Inc., Boca Raton, Fla.

Peuquet, D. J. and Marble, D. E. 1990. *Introductory Readings in Geographic Information Systems*. Taylor and Francis, London.

Pike, R. J. 1988. "The Geometric Signature: Quantifying Landslide Susceptible Terrain Types from Digital Elevation Models." *Mathematical Geology*, vol. 20, pp. 491-511.

Preparata, F. P. and Shamos, M. I. 1985. *Computational Geometry: An Introduction*. Springer-Verlag, New York.

Richardus, P. and Adler, R. K. 1972. *Map Projections for Geodisists, Cartographers and Geographers*.

Rogers, E. M. 1962. *Diffusion of Innovations*. Free Press, New York.

Rosenblum, L., Earnshaw, R.A., Encarnacao, J., Hagen, H., Kaufman, A., Klimenko, S., Nielson, G., Post, F. and Thalmann, D. 1994. *Scientific Visualization: Advances and Challenges*. Academic Press, New York.

Salem, L., Testard, F. and Salem, C. 1992. *The Most Beautiful Mathematical Formulas*. John Wiley, New York.

Sampson, R. J. 1978. *Surface II Graphics System*. Series on Spatial Analysis, #1, Kansas Geological Survey, Lawrence, Kan.

Sampson, R. J. 1994. *Surface III Manual*. Surface III Office, Kansas Geological Survey, Lawrence, Kan.

Siegel, S. and Castellan, N. J. 1988. *Nonparametric Statistics for the Behavioral Sciences* (Second Edition). McGraw-Hill, New York.

Simon, H. 1991. *Models of My Life*. Basic Books, New York.

Snapper, E. and Troyer, R. J. 1971. *Metric Affine Geometry*. Academic Press, New York.

Snyder, J. P. 1987. *Map Projections—A Working Manual*. U.S. Geological Survey Professional Paper 1,395. U.S. Government Printing Office, Washington, D.C.

Tabachnik, B. G. and Fidell, L. S. 1989. *Using Multivariate Statistics*. Harper and Row, New York.

Taylor, P. J. 1975. "Distance Decay in Spatial Interactions." *Concepts and Techniques in Modern Geography, #2*, Geo Abstracts Ltd., Norwich, England.

Waters, N.M. 1989. Geography, "Microcomputers and the Use of Color." *The Operational Geographer*, vol. 7 #3, pp. 33-38.

Waters, N. M. 1994a. "The Geographic Calculator." *GIS WORLD*, vol. 6, # 5, p. 64.

Waters, N. M. 1994b. "Statistics: How Much Should the GIS Analyst Know?" *GIS WORLD*, vol. 6, # 3, p. 62.

Weibel, R. and Heller, M. 1991. "Digital Terrain Modeling." Chapter 19, pp. 269-297, in Maguire et al., op. cit.

Wellar, B. and Wilson, P. 1993. "Contributions of GIS Concepts and Capabilities to Scientific Inquiry: Initial Findings." Pp. 753-767 in the *Proceedings of GIS/LIS '93*, Association of American Geographers, Washington, D.C.

Williams, R. B. G. 1984. *Introduction to Statistics for Geographers and Earth Scientists*. MacMillan Publishers Ltd., London.

Wilson, A. G. and Kirkby, M. J. 1980. *Mathematics for Geographers and Planners*. Clarendon Press, Oxford.

Resources

Professional Associations

American Congress on Surveying and Mapping (ACSM)
5410 Grosvenor Lane, Bethesda, MD 20814
301-493-0200; fax: 301-493-8245

The national association for professionals in surveying, cartography, geodesy, GIS, land information systems (LIS) and related fields, ACSM comprises several member organizations: the Geographic and Land Information Society (GLIS), the American Cartographic Association (ACA), the American Association for Geodetic Surveying (AAGS) and the National Society for Professional Surveyors (NSPS). The association has eight regional sections or chapters and 28 student chapters. Among its objectives are the improvement of college curricula for teaching all branches of surveying, cartography, geodesy, GIS/LIS and affiliated fields in the technological sciences and the professional philosophies. ACSM offers its members numerous benefits including news, technical, and scholarly publications; education and certification programs; fellowships and scholarships; legislative-affairs action; and annual conferences. Its publications include the *ACSM Bulletin*, a bimonthly news magazine, and two scholarly journals: *Surveying and Land Information Systems and Cartography* and *Geographic Information Systems*.

American Society for Photogrammetry and Remote Sensing (ASPRS)
5410 Grosvenor Lane, Bethesda, MD 20814
301-493-0290; fax: 301-493-0208

ASPRS, a high-technology society serving GIS professionals, began as the American Society of Photogrammetry in 1934, and increased its service to the scientific community as the mapping sciences evolved and expanded. ASPRS is divided into five thematic divisions: Geographic Information Systems, Remote Sensing Applications, Photogrammetric Applications, Primary Data Acquisition and Professional Practice. Members may be part of any or all divisions. The primary purpose of ASPRS is to disseminate information, and the main pipeline to members is the monthly journal *Photogrammetric Engineering & Remote Sensing* (PE&RS). ASPRS also has an extensive catalog of books, videos and films which are available at a substantial discount to members.

Asociación Española de Sistemas de Información Geográfica (AESIG)
Real Academia de Ciencias, C/Valverde 22
28004 Madrid, SPAIN
34-1-441-7799; fax: 34-1-442-4889

The Spanish GIS Association was formed in 1990 to create and maintain a cohesive GIS community in Spain. Its main goal, as an organization that is independent of government agencies and system vendors, is internal technology transfer, primarily information dissemination. AESIG publishes a quarterly bulletin to help keep the Spanish GIS community abreast of national, European and international news regarding GIS and geographic information policy. It offers seminars on specific topics such as research and development trends, funding opportunities and data availability and participates in organizing the major annual GIS conference in Spain.

The Association for Geographic Information (AGI)
12 Great George Street, London SW1P 3AD UK
44-71-222-7000 ext. 226; fax 44-71-222-9430

A hybrid of learned society and trade association, with 650 members, roughly half individuals and half corporate organizations, the AGI represents both suppliers and users of geographic information systems and data. It has both private and public sector members including local authorities, central government departments and government agencies. AGI sponsors publications, conferences and seminars; liaises with government departments and agencies and with other organizations; and undertakes projects such as the development of standards in the United Kingdom and Europe.

Association of American Geographers (AAG)
1710 Sixteenth St. NW, Washington, D.C. 20009-3198
202-234-1450; fax: 202-234-2744

AAG, founded in 1904, promotes and encourages geographic research and education and disseminates research findings. Its members include teachers, business and government employees, students, and independent scholars in the United States, Canada and other countries. As geographers, they analyze and explain where things are and why they are there, seek explanations for settlement and migration patterns, study major regions of the world and explore the ways humankind uses and abuses its physical environments. Its annual meetings focus on regions and topical specialties such as environmental studies, national and human-induced hazards, population and migration, cartography, geopolitics, planning and GIS. The AAG also sponsors a GIS/LIS conference each year in cooperation with the ACSM, AM/FM, ASPRS and URISA.

The Association of Chinese Professionals in Geographic Information Systems - Abroad (CPGIS)
Chinese University of Hong Kong, Geography Department
Shatin, NT, HONG KONG
http://umgis.merrick.miami.edu/bli-html/cpgis/cpgis.html

Founded in 1992 by a group of Chinese professionals and graduate students, CPGIS now has more than 360 members in GIS, remote sensing, GPS and other GIS-related fields. The Board of Directors meets monthly through the electronic mail (e-mail) network. Several committees focus on academic and business development, information sharing and exchange, public relations, etc. CPGIS has an e-mail network, CPGIS-L, to connect its members in Asia, Australia, Europe and North America. The association sponsors an annual conference; organizes GIS training courses, workshops and conferences in China; participates in coordinating and editing the Chinese GIS Handbook; provides a guided tour service to some academic/government delegations from China; and assists scholars from other countries in China. It also publishes a bimonthly newsletter on GIS development in China and other regions and a new journal on recent research results and successful applications.

The Australasian Urban and Regional Information Systems Association, Inc. (AURISA)
PO Box E307, Queen Victoria Terrace
ACT 2600 AUSTRALIA
616-273-4054; fax: 616-273-4057

AURISA was established in 1975 to promote awareness and advance the development of urban and regional information systems; provide a forum for the exchange of information about the technology in Australia, New Zealand and the Asia-Pacific region; and to address the public policy issues emerging in this field. Its multidisciplinary membership is drawn from local, state and federal government authorities, public utilities, academia, vendors, and consultants with an interest in the development and application of urban and regional information systems which include GIS. The AURISA Annual Conference provides an opportunity for delegates to inspect a technical exposition of the latest developments in hardware and software and to participate in a number of group sessions and technical presentations as well as workshops. AURISA produces a number of publications, including the *AURISA News*, technical monographs, and the proceedings of conferences and seminars.

Automated Mapping/Facilities Management International (AM/FM)
14456 East Evans Avenue, Aurora, CO 80014
303-337-0513; fax: 303-337-1001

AM/FM International fosters information exchange, educational opportunities and scientific research and development to advance the benefits of geographic and facilities management information systems. It serves utilities; local, state and federal government agencies; other interested organizations; and the general public, providing educators, experts and novices with a forum for discussing processes, problems and solutions to accessing and managing mapping and facilities database information. Its annual conference features seminars and presentations to meet members' educational needs. The association's bimonthly newsletter, *AM/FM/GIS Networks*, is distributed to all members. AM/FM International offers annual scholarships and internships to selected graduate and undergraduate students enrolled in related higher education degree programs.

Canadian Association of Geographers (CAG)
Burnside Hall, McGill University
Rue Sherbrooke St. W, Montreal, Quebec H3A 2K6, CANADA
514-398-4946; fax: 514-398-7437

Founded in 1951, CAG is the collective voice of Canadian geography in Canada and abroad. The association, whose members include geographers in many walks of life, promotes geographical research and teaching and represents the profession in the scientific and business communities. Through publications and conferences, CAG provides a forum for the exchange of ideas and the dissemination of geographical research and information concerning recent theoretical, technical and applied developments in the field. Membership is open to geographers engaged in research, education, business and government and to others, including students, who share the association's objectives. The "flagship" publication, *The Canadian Geographer*, serves the entire field of geography, both physical and human, and is circulated to all members. The *CAG Newsletter* is published six times a year for members and subscribers. The association holds an annual conference on specific geography aspects, with special sessions devoted to issues of interest to the general public, such as environmental concerns, global change, water resources and sustainable development.

Geographic Information Systems Association (GISA)
Okabe Lab, Department of Urban Engineering
University of Tokyo, 7-3-1 Hongo, Bunkyo-ku
Tokyo 113 , JAPAN
813-5800-6964; fax: 813-5800-6964

An academic society for promoting the development of the theory and applications of GIS, GISA was founded in 1991. The members include researchers at universities (almost half), institutes, municipal governments, surveying companies, consultants, electronic/computer companies and students. GISA holds its own annual conference and cosponsors the annual Functional Graphic Information Systems (multimedia with GIS) conference organized by the Institute of Electronics, Information and Communication Engineers. GISA has eight special interest groups: functional geographic information systems, forest planning, forest landscape planning with GIS (Forest View), business GIS, GIS in local governments, object-oriented GIS, disaster prevention planning with GIS, and GIS terminology and GIS education. GISA publishes a refereed journal, *Theory and Applications of GIS*, and a newsletter.

Urban and Regional Information Systems Association (URISA)
900 Second St. NE, Suite 304, Washington, D.C. 20002
202-289-1685; fax: 202-842-1850

URISA is a professional/educational organization for individuals concerned with information systems in local, regional and state/provincial governments. Founded in 1963, it supports the belief that information systems technology can and will improve decision making by public officials. Its membership represents a multi-disciplinary cross-section of government, private industry and academic professionals which brings together information users and providers, managers and technicians, and analysts and vendors. URISA holds an annual summer conference with pre-conference workshops to cover introductory and advanced topics such as GIS, permit tracking and infrastructure management. It also produces a number of publications including the semi-annual *URISA Journal* and sponsors a contest for Exemplary Systems in Government. URISA has chapters throughout the United States and Canada and special interest groups which focus on GIS; infrastructure management; integrated systems; urban and regional analysis; environment and natural resources; transportation; education and technology transfer; land records modernization; public safety; regional agencies; public information; spatial decision support systems; artificial intelligence/knowledge-based systems; multmedia and water/waste water utilities.

Information Clearinghouses

National Center for Geographic Information and Analysis (NCGIA)
University of California at Santa Barbara, 3510 Phelps Hall
Santa Barbara CA 93106-4060
805-893-8224; fax 805-893-8617

A consortium of three universities—University of California at Santa Barbara, University of Maine at Orono, and the State University of New York at Buffalo—NCGIA endorses standardizing guidelines for GIS curricula and maintains a library and bibliography of articles and publications on GIS and GIS-related issues. It sponsors a number of conferences and technical workshops.

U.S. Geological Survey (USGS)
Earth Science Information Center (ESIC)
507 National Center, Reston VA 22092
713-646-6045

A good source of digital line graph (DLG) data collected primarily by federal agencies. Call 800-USA-MAPS to contact any of the regional ESICs.

Condensed from *GIS World Sourcebook 1996*, Fort Collins, CO, GIS World, Inc., forthcoming 1995.

Companion Software

Two software systems support the material presented in *Spatial Reasoning*: a set of digital slide shows that demonstrates GIS concepts and a set of hands-on tutorials that provides practical experience. The GIS Concepts (gCON) slide shows give thorough graphic explanations of many of the concepts presented in this book. The Tutorial Map Analysis Package (tMAP) is an easy way to get your hands on GIS technology and gain experience in the application of map analysis concepts. The tMAP and gCON systems also support the companion book, *Beyond Mapping: Concepts, Algorithms, and Issues in GIS* (GIS World Books, 1993), and are internally cross-referenced to the topics it covers. The following text describes both the gCON and tMAP systems and cross-references them to *Spatial Reasoning*. To order *Beyond Mapping*, tMAP, or gCON, use the order form at the back of this book or contact

GIS WORLD BOOKS
155 E. Boardwalk Drive, Suite 250
Fort Collins, CO 80525
phone: 970-223-4848 • fax: 970-223-5700
E-mail: books@gisworld.com

The GIS Concepts Digital Slide Shows (gCON™)

The gCON system presents map analysis concepts for self-learning through slide sets that demonstrate GIS procedures and applications. It contains 19 digital slide shows that extend many of the figures in *Spatial Reasoning*.

The gCON slides were developed during the last 20 years by Joseph K. Berry for a variety of university courses and professional workshops. They are selected from the GIS Digital Slide Show Series distributed by Berry & Associates, 2000 S. College Ave., Suite 300, Fort Collins, CO 80525 [phone: 970-490-2155].

gCON Specifications

- *Disk Format:* Two 3.5", DS/HD, DOS formatted (1.44MB) diskettes

- *Environment:* Any personal computer (PC) with a hard disk, EGA or higher color graphics, and DOS Version 3.2 or higher operating system; may be accessed as DOS application under Windows 3.1 or higher; Intel '286, '386, '486 and Pentium central processors.

- *Presentation Software*: Graphic files are in .PIC format and are accessed using the Presents Slide Show Utility by Media Cybernetics, Inc. under non-exclusive distribution license to Berry & Associates.

- gCON Version 1.1 copyright ©1995 by Berry & Associates. All Rights Reserved. Licensed for personal, non-commercial use only.

gCON Digital Slide Shows

The following list groups the 19 slide shows by their content and cross references them to the topics in *Spatial Reasoning*.

Basic Concepts

OVIEW.EXE	Introduction to computer mapping, management and analysis
DBCONS.EXE	Database concepts
DTERM.EXE	Data structure terminology
DSTRUC.EXE	Data structure formats and relative advantages
CONS.EXE	General considerations in GIS

Analytical Concepts

FUNCT.EXE	Outline of the fundamental classes of analytic GIS functions
STAT.EXE	Treating maps as data using spatial stat/math
SSTAT.EXE	Iterative smoothing interpolation technique
DIST.EXE	Effective distance measurement
VIEW.EXE	Visual analysis
SHAPE.EXE	Quantitative measures for assessing feature shape

Modeling Concepts

HUMANE.EXE	A procedure to link model logic to GIS commands
HONEST.EXE	Mapping uncertainty and error propagation
LSLIDE.EXE	Landslide susceptibility demonstrating binary, rating, weighted average, and extended models
RUSLE.EXE	Soil loss model demonstrating a disaggregated spatial model

Applications

BB-BK.EXE	Conflict resolution model of island planning for conservation, research and development
SED.EXE	Sediment loading potential model considering slope and cover for variable-width buffers
F-RESP.EXE	Wildfire response model based on both on- and off-road conditions
F-RISK.EXE	Wildfire risk model based on fuel loading, detection and response time

Spatial Reasoning *Topic/gCON Cross Reference*

Introduction

OVIEW.EXE	Introduction to computer mapping, management, and analysis

Topic 1

HUMANE.EXE	A procedure to link model logic to GIS commands
BB-BK.EXE	Conflict resolution model of island planning for conservation, research and development

Topic 2

STAT.EXE	Introduction to treating maps as data using spatial stat/math
SSTAT.EXE	Iterative smoothing interpolation technique

Topic 3
DTERM.EXE   Data structure terminology
CONS.EXE   General considerations in GIS

Topic 4
HONEST.EXE   Mapping uncertainty and error propagation

Topic 5
FUNCT.EXE   Fundamental classes of analytic GIS functions

Topic 6
DSTRUC.EXE   Data structure formats and relative advantages

Topic 7
DBCONS.EXE   Database concepts
DIST.EXE   Effective distance measurement
VIEW.EXE   Visual analysis

Topic 8
LSLIDE.EXE   Landslide susceptibility demonstrating binary, rating, weighted
                 average, and extended models
RUSLE.EXE   Soil loss model demonstrating a disaggregated, spatial model

Topic 9
SED.EXE   Sediment loading potential model considering slope and cover
                 for variable-width buffers
F-RESP.EXE   Wildfire response model based on both on- and off-road conditions
F-RISK.EXE   Wildfire risk model based on fuel loading, detection and
                 response time

Topic 10
SHAPE.EXE   Quantitative measures for assessing feature shape

The Tutorial Map Analysis Package (tMAP™)

The tMAP system provides self-learning of map analysis concepts through hands-on experience. There are 20 tutorials and 13 application models cross-referenced to the topics presented in *Spatial Reasoning*. In addition, users are shown how to encode their own data and relate their own application models.

The tMAP system is a special version of the Professional Map Analysis Package (pMAP™) commercial software and the Academic Map Analysis Package (aMAP™) educational materials for classroom study. For more information about pMAP and aMAP contact Spatial Information Systems Inc., 2000 S. College Ave., Suite 300, Fort Collins, CO 80525 [phone: 970-490-2155].

The tutorial package contains fully functional software, a tutorial database, exercises and brief text. Within a user-friendly, natural language, tMAP provides advanced map processing analysis capabilities, including:

optimal paths • visual exposure • edge characterization • spatial interpolation • slope/aspect • proximity • coincidence statistics • roving window summaries • map overlay • geographic searches • thematic mapping • contouring • shape/pattern analysis • contiguity

The tutorials are accessed through an easy-to-use menu interface. The tMAP system uses a raster data structure, yet allows for input in the form of gridded data, digitized points, lines or polygons. Output includes summary tables, character-based or raster pattern maps, monochrome or color display, and standard files for enhanced color and line graphics.

tMAP Specifications

- *Disk Format:* One 3.5", DS/HD, DOS formatted (1.44MB) diskette

- *Environment:* Any personal computer (PC) with a hard disk and DOS Version 3.2 or higher operating system; may be accessed as DOS application under Windows 3.1 or higher; Intel 286, 386, 486 and Pentium central processors; math coprocessor recommended but not required.

- *Map Size:* 25 rows by 25 columns (Tutorial Database) 80 rows by 80 columns (maximum configuration)

- tMAP Version 3.12 copyright ©1995 by Spatial Information Systems, Inc. All Rights Reserved. Licensed for personal, non-commercial use only.

tMAP Tutorials

The following list groups the 20 tutorials and 13 models by degree of difficulty and cross references them to the topics in *Spatial Reasoning*.

Beginning to Intermediate

TUTOR1-9.CMD   Tutorials corresponding to the brief text (TEXT. DOC) on the tMAP diskette.

Intermediate to Advanced

TMAP0-10.CMD   Tutorials demonstrating analytic capabilities as described in Beyond Mapping

Advanced

TU-ERODE.CMD   Simple erosion potential model
TU-ACT.CMD   Human activity derivation model
TU-VIEW.CMD   Visual exposure analysis model
TU-INTRP.CMD   Spatial interpolation model
TU-SED.CMD   Effective distance sediment loading model
TU-SLIDE.CMD   Landslide susceptibility model
TU-ACCES.CMD   Timber access and facilities siting model
TU-RESP.CMD   Wildfire response model (on- and off-roads)
TU-RISK.CMD   Wildfire risk model (fuel, detection and response)
TU-WATER.CMD   Encoding points, lines, and polygons
TU-POLY.CMD   Automatic polygon fill and tagging
TU-SOILS.CMD   Encoding adjacent polygons
TU-ELEV.CMD   Encoding elevation

Spatial Reasoning Topic/tMAP Cross Reference

Introduction

TMAP0.CMD   Basic introduction to tMAP system
TUTOR9.CMD   Data entry into tMAP

Topic 1

TMAP10.CMD	Campground development model
TUTOR2.CMD	Reclassifying maps
TUTOR3.CMD	Point-by-point overlay

Topic 2

TMAP1.CMD	Spatial interpolation
TU-INTRP.CMD	Spatial interpolation model
TU-ELEV.CMD	Considerations in encoding elevation

Topic 3

TUTOR4.CMD	Region-wide overlay
TMAP7.CMD	Overlaying maps
TU-ACT.CMD	Human activity derivation model

Topic 4

TMAP6.CMD	Uncertainty and error propagation

Topic 5

TUTOR5.CMD	Distance measurement
TUTOR6.CMD	Connectivity operations
TUTOR7.CMD	Neighborhood summaries
TU-VIEW.CMD	Visual exposure analysis model

Topic 6

TU-WATER.CMD	Encoding points, lines and polygons
TU-SOILS.CMD	Encoding adjacent polygons
TU-POLY.CMD	Automatic polygon fill and tagging

Topic 7

TMAP2.CMD	Effective distance
TMAP3.CMD	Roving windows
TMAP9.CMD	More on slope distance and viewsheds

Topic 8

TMAP8.CMD	Analyzing powerline siting
TU-SLIDE.CMD	Landslide susceptibility model
TU-ERODE.CMD	Simple erosion potential model

Topic 9

TU-SED.CMD	Effective distance sediment loading model
TU-RESP.CMD	Wildfire response model
TU-RISK.CMD	Wildfire risk model
TU-ACCES.CMD	Timber access and facilities siting model

Topic 10

TMAP4.CMD	Potpourri of map analysis operations
TMAP5.CMD	Assessing shape and pattern

Index

ORDER FORM

TITLE	PRICE	AMOUNT
Special *with this order form only,* ☐ gCON software	$13.80	
☐ Beyond Mapping (BMAP) ☐ Spatial Reasoning (SR)	$32.95 each	
☐ Both Books	$55	
☐ tMAP software	$21.95 each	
Either book ☐ BMAP ☐ SR and either disk ☐ tMAP ☐ gCON	$46.75	
Either book ☐ BMAP ☐ SR and both sets of software	$63	
Both books and either ☐ tMAP or ☐ gCON	$78	
☐ Both books and both sets of software	$93	

	Subtotal	
Colorado residents add 3.25% tax; Canadian residents add 7% GST.		
Shipping and handling: U.S. orders add 10%; Canadian orders add 15%; elsewhere add 20%.		
	Total	

T o   o r d e r :
1. **Phone:** 970-223-4848
2. **FAX:** 970-223-5700
3. **Mail:** GIS World Inc. 155 E. Boardwalk Drive, Suite 250, Fort Collins, CO 80525, USA
4. **E-mail:** books@gisworld.com

Name_____

Organization _____

Address_____

_____

City _____

State/Province_____

ZIP/Mail Code _____Country _____

Phone_____Fax _____

E-mail _____

☐ VISA   ☐ MasterCard   ☐ Check/Money Order

☐ Purchase Order no _____

Account Number:

☐☐☐☐☐☐☐☐☐☐☐☐☐☐☐☐☐☐☐☐

Signature _____Expiration Date _____

97SR01